应用型本科院校"十二五"规划教材/计算机类

主　编　唐友　舒杰
副主编　丁龙　张鑫　陈中星　赵鑫
主　审　葛雷

Java 语言程序设计实验指导

Experimental Guidance for Java Programming

哈尔滨工业大学出版社

内 容 简 介

全书共分 17 章：前 16 章中"典型例题解析"和"课后习题解答"是典型例题和《Java 语言程序设计》一书中全部课后习题的详细分析、解答及程序上机运行结果，每章中"上机实验"是精心设计的实验及相应的程序代码；第 17 章"综合实例"包括两个具有较高综合性的编程实例。

图书在版编目(CIP)数据

Java 语言程序设计实验指导/唐友,舒杰主编.—哈尔滨：哈尔滨工业大学出版社,2013.1

应用型本科院校"十二五"规划教材

ISBN 978-7-5603-3522-3

Ⅰ.①J… Ⅱ.①唐…②舒… Ⅲ.①JAVA 语言-程序设计-高等学校-教学参考资料 Ⅳ.①TP312

中国版本图书馆 CIP 数据核字(2012)第 278910 号

策划编辑	赵文斌 杜 燕
责任编辑	李广鑫
出版发行	哈尔滨工业大学出版社
社 址	哈尔滨市南岗区复华四道街 10 号 邮编 150006
传 真	0451-86414749
网 址	http://hitpress.hit.edu.cn
印 刷	黑龙江省地质测绘印制中心印刷厂
开 本	787mm×1092mm 1/16 印张 19.75 字数 500 千字
版 次	2013 年 1 月第 1 版 2013 年 1 月第 1 次印刷
书 号	ISBN 978-7-5603-3522-3
定 价	35.80 元

(如因印装质量问题影响阅读,我社负责调换)

《应用型本科院校"十二五"规划教材》编委会

主　任　修朋月　竺培国
副主任　王玉文　吕其诚　线恒录　李敬来
委　员　（按姓氏笔画排序）
　　　　　丁福庆　于长福　马志民　王庄严　王建华
　　　　　王德章　刘金祺　刘宝华　刘通学　刘福荣
　　　　　关晓冬　李云波　杨玉顺　吴知丰　张幸刚
　　　　　陈江波　林　艳　林文华　周方圆　姜思政
　　　　　庹　莉　韩毓洁　臧玉英

序

哈尔滨工业大学出版社策划的《应用型本科院校"十二五"规划教材》即将付梓，诚可贺也。

该系列教材卷帙浩繁，凡百余种，涉及众多学科门类，定位准确，内容新颖，体系完整，实用性强，突出实践能力培养。不仅便于教师教学和学生学习，而且满足就业市场对应用型人才的迫切需求。

应用型本科院校的人才培养目标是面对现代社会生产、建设、管理、服务等一线岗位，培养能直接从事实际工作、解决具体问题、维持工作有效运行的高等应用型人才。应用型本科与研究型本科和高职高专院校在人才培养上有着明显的区别，其培养的人才特征是：①就业导向与社会需求高度吻合；②扎实的理论基础和过硬的实践能力紧密结合；③具备良好的人文素质和科学技术素质；④富于面对职业应用的创新精神。因此，应用型本科院校只有着力培养"进入角色快、业务水平高、动手能力强、综合素质好"的人才，才能在激烈的就业市场竞争中站稳脚跟。

目前国内应用型本科院校所采用的教材往往只是对理论性较强的本科院校教材的简单删减，针对性、应用性不够突出，因材施教的目的难以达到。因此亟须既有一定的理论深度又注重实践能力培养的系列教材，以满足应用型本科院校教学目标、培养方向和办学特色的需要。

哈尔滨工业大学出版社出版的《应用型本科院校"十二五"规划教材》，在选题设计思路上认真贯彻教育部关于培养适应地方、区域经济和社会发展需要的"本科应用型高级专门人才"精神，根据黑龙江省委书记吉炳轩同志提出的关于加强应用型本科院校建设的意见，在应用型本科试点院校成功经验总结的基础上，特邀请黑龙江省9所知名的应用型本科院校的专家、学者联合编写。

本系列教材突出与办学定位、教学目标的一致性和适应性，既严格遵照学科体系的知识构成和教材编写的一般规律，又针对应用型本科人才培养目标及与之相适应的教学特点,精心设计写作体例,科学安排知识内容,围绕应用讲授理论,做到"基础知识够用、实践技能实用、专业理论管用"。同时注意适当融入新理论、新技术、新工艺、新成果,并且制作了与本书配套的PPT多媒体教学课件,形成立体化教材,供教师参考使用。

《应用型本科院校"十二五"规划教材》的编辑出版,是适应"科教兴国"战略对复合型、应用型人才的需求,是推动相对滞后的应用型本科院校教材建设的一种有益尝试,在应用型创新人才培养方面是一件具有开创意义的工作,为应用型人才的培养提供了及时、可靠、坚实的保证。

希望本系列教材在使用过程中,通过编者、作者和读者的共同努力,厚积薄发、推陈出新、细上加细、精益求精,不断丰富、不断完善、不断创新,力争成为同类教材中的精品。

黑龙江省教育厅厅长

前　言

　　Java语言是一门发展非常快、不断创新的计算机语言，许多大型项目已经采用了Java语言来开发。"Java程序设计"课程得到许多学校的关注。为了适应计算机教学发展趋势，有必要编写一本符合当前Java语言发展趋势和教学现状的实验指导，帮助广大读者了解和掌握Java语言的当前的特点，以克服读者"学"和"实用"脱节的问题。

　　结合编者从事Java课程教学活动中积累的经验，从实用性、科学性以及当前的计算机技术出发编写本书。

　　为了让读者更好地学习Java语言，《Java语言程序设计实验指导》将每章分成典型例题、课后习题解答、实验指导及程序代码几个部分。典型例题部分对学习过程中需要注意的知识点和一些常见的问题做了归纳和总结，能帮助读者对关键知识点快速地了解和巩固。实验指导部分由浅入深，通过详细的实验步骤和完整的实验设计指导每个实验，通过程序改错、补充程序、程序分析、独立编写程序，以及问题思考等多种方法，立体地指导读者来深入理解和掌握Java语言，克服了传统实验指导中存在实验手段单一的问题，较好地通过实验来学习和掌握理论知识。

　　这是一本针对学习Java语言的实验指导教材。本书分成17章，涵盖了当前J2SE中的初级、中级大部分内容和高级编程技术的部分内容，包括当前Java的主流Eclipse开发环境，Java的基本数据类型和基本运算，Java控制语句，数组，类和对象，包和接口的应用，异常处理，字符串处理，Applet小应用程序，Java的GUI编程，事件处理Java的文件处理，Java的网络技术的实现以及Java的数据库初步编程，以及最后提供两个完整的综合案例。

　　本教材作为黑龙江省高等教育教学改革项目《基于应用型本科院校创新型卓越软件人才培养模式的研究》研究成果之一。

　　本教材由唐友、舒杰任主编；丁龙、张鑫、陈中星、赵鑫任副主编；陈瑶、田崇瑞、耿姝、张珑、郭鑫、杨迎、刘荣军、刘立栋、单晓光、贾仁山、赵丹、董晶、车玉生参编。作者编写分工如下：第1、2章由田崇瑞编写，第3章由耿姝编写，第4、10章和第17章的7.2节由唐友编写，第5章由陈瑶编写，第6章由陈中星编写，第7、8、9章和第17章的7.1节由丁龙编写，第11、12、14章由舒杰编写，第13章由张鑫编写，第15章由赵鑫编写，第16章由刘荣军、张珑、郭鑫、杨迎、刘立栋、单晓光、贾仁山、赵丹、董晶、车玉生编写。本教材编写还得到了哈尔滨德强商务学院、哈尔滨师范大学、齐齐哈尔大学、哈尔滨广厦学院、黑龙江东方学院、哈尔滨石油学院、哈尔滨华德学院、东北农业大学成栋学院、黑龙江生物科技职业学院等院校及哈尔滨晨星科技开发有限公司有关领导的大力支持，在此深表谢意。全书在葛雷副教授的主审下完成。

　　由于编者水平，虽经努力，教材一定仍存有各种问题，恳请广大读者提出宝贵意见和建议，以便修订时加以完善。

编　者
2013年1月

目 录

第1章 Java 语言概述 ································· 1
   1.1 典型例题解析 ································· 1
   1.2 课后习题解答 ································· 1
   1.3 上机实验 ······································ 3
   1.4 程序代码 ······································ 4

第2章 Java 语言基础 ································ 12
   2.1 典型例题解析 ································ 12
   2.2 课后习题解答 ································ 16
   2.3 上机实验 ····································· 18
   2.4 程序代码 ····································· 18

第3章 基本控制结构 ································· 21
   3.1 典型例题解析 ································ 21
   3.2 课后习题解答 ································ 23
   3.3 上机实验 ····································· 28
   3.4 程序代码 ····································· 29

第4章 数组、方法与字符串 ······················· 33
   4.1 典型例题解析 ································ 33
   4.2 课后习题解答 ································ 38
   4.3 上机实验 ····································· 45
   4.4 程序代码 ····································· 46

第5章 类和对象 ·· 49
   5.1 典型例题解析 ································ 49
   5.2 课后习题解答 ································ 52
   5.3 上机实验 ····································· 57
   5.4 程序代码 ····································· 57

第6章 类的继承和多态 ······························ 61
   6.1 典型例题解析 ································ 61
   6.2 课后习题解答 ································ 68
   6.3 上机实验 ····································· 72
   6.4 程序代码 ····································· 73

第7章 接口、抽象类与包 ··························· 77
   7.1 典型例题解析 ································ 77
   7.2 课后习题解答 ································ 80
   7.3 上机实验 ····································· 87

7.4 程序代码 ………………………………………………………………………………… 88

第 8 章 异常处理 …………………………………………………………………………… 94
8.1 典型例题解析 …………………………………………………………………………… 94
8.2 课后习题解答 …………………………………………………………………………… 103
8.3 上机实验 ………………………………………………………………………………… 107
9.4 程序代码 ………………………………………………………………………………… 107

第 9 章 集合类 ……………………………………………………………………………… 110
9.1 典型例题解析 …………………………………………………………………………… 110
9.2 课后习题解答 …………………………………………………………………………… 115
9.3 上机实验 ………………………………………………………………………………… 117
9.4 程序代码 ………………………………………………………………………………… 118

第 10 章 多线程 …………………………………………………………………………… 125
10.1 典型例题解析 ………………………………………………………………………… 125
10.2 课后习题解答 ………………………………………………………………………… 130
10.3 上机实验 ……………………………………………………………………………… 134
10.4 程序代码 ……………………………………………………………………………… 135

第 11 章 图形用户界面设计 ……………………………………………………………… 142
11.1 典型例题解析 ………………………………………………………………………… 142
11.2 课后习题解答 ………………………………………………………………………… 148
11.3 上机实验 ……………………………………………………………………………… 158
11.4 程序代码 ……………………………………………………………………………… 159

第 12 章 Swing 组件 ……………………………………………………………………… 163
12.1 典型例题解析 ………………………………………………………………………… 163
12.2 课后习题解答 ………………………………………………………………………… 168
12.3 上机实验 ……………………………………………………………………………… 180
12.4 程序代码 ……………………………………………………………………………… 180

第 13 章 Applet 程序 ……………………………………………………………………… 185
13.1 典型例题解析 ………………………………………………………………………… 185
13.2 课后习题解答 ………………………………………………………………………… 188
13.3 上机实验 ……………………………………………………………………………… 192
13.4 程序代码 ……………………………………………………………………………… 192

第 14 章 输入与输出 ……………………………………………………………………… 197
14.1 典型例题解析 ………………………………………………………………………… 197
14.2 课后习题解答 ………………………………………………………………………… 207
14.3 上机实验 ……………………………………………………………………………… 211
14.4 程序代码 ……………………………………………………………………………… 211

第 15 章 数据库编程 ……………………………………………………………………… 217
15.1 典型例题解析 ………………………………………………………………………… 217
15.2 课后习题解答 ………………………………………………………………………… 222
15.3 上机实验 ……………………………………………………………………………… 225

  15.4 程序代码 ·· 226
第 16 章 网络程序设计 ·· 232
  16.1 典型例题解析 ··· 232
  16.2 课后习题解答 ··· 236
  16.3 上机实验 ·· 239
  16.4 程序代码 ·· 239
第 17 章 综合案例 ··· 245
  17.1 蜘蛛纸牌 ·· 245
  17.2 Java 聊天室 ··· 262
参考文献 ··· 302

第 1 章

Java 语言概述

1.1 典型例题解析

1. 编写 Java 源程序

打开一个纯文本编辑器,键入如下程序:
```
public class Hello {
    public static void main(Stringargs[ ]) {
        System. out. println("Hello,welcome!");
    }
}
```
将文件命名为 Hello. java,保存为文本文件,注意保存文件的路径。根据前面环境变量的设置,Hello. java 应该保存在"e:\java\程序"的路径下。

2. 编译 Java 源程序

Java 源程序编写后,要使用 Java 编译器(javac. exe)进行编译,将 Java 源程序编译成可执行的程序代码。Java 源程序都是扩展名为. java 的文本文件。编译时首先读入 Java 源程序,然后进行语法检查,如果出现问题就终止编译。语法检查通过后,生成可执行程序代码即字节码,字节码文件名和源文件名相同,扩展名为. class。

打开命令提示符窗口或 MS-DOS 窗口进入 Java 源程序所在路径。

键入编译器文件名和要编译的源程序文件名 javac Hello. java。

按回车键开始编译(注意:文件名 H 要大写,否则运行会出问题)。

如果源程序没有错误,则屏幕上没有输出,键入"dir"按回车键后可在目录中看到生成了一个同名字的. class 文件"Hello. class"。否则,将显示出错信息。

3. 运行 Java 程序

使用 Java 解释器(java. exe)可将编译后的字节码文件 Hello. class 解释为本地计算机代码。在命令提示符窗口或 MS-DOS 窗口键入解释器文件名和要解释的字节码文件名 javaHello,按回车键即开始解释并可看到运行结果,表明程序运行成功了。

1.2 课后习题解答

1. 简述面向对象软件开发方法的重要意义。

答：面向对象的软件开发方法按问题论域来设计模块，以对象代表问题解的中心环节，力求符合人们日常的思维习惯，采用"对象+消息"的程序设计模式，降低或分解问题的难度和复杂性，从而以较小的代价和较高的收益获得满意的效果，满足软件工程发展的需要。

2. 解释下面几个概念。

（1）对象　（2）实例　（3）类　（4）消息　（5）封装　（6）继承　（7）多态

答：（1）对象：现实世界中某个具体的物理实体在计算机中的映射和体现。它包括属性和作用于属性的操作。

（2）实例：对象在计算机内存中的映像称为实例。

（3）类：类是面向对象技术中一个非常重要的概念，它是描述对象的"基本原型"，是描述性的类别或模板，即对一组对象的抽象。

（4）消息：对象的动作取决于外界给对象的刺激，这就是消息，即消息是对象之间进行通信的一种数据结构。

（5）封装：所谓封装又称为信息隐蔽，是面向对象的基本特征。

（6）继承：继承是面向对象语言中的一种重要机制，该机制自动地为一个类提供来自另一个类的操作和属性。

（7）多态：多态是指一个名字具有多种语义，即指同一消息为不同对象所接受时，可以导致不同的操作。

3. 对象"狗"与对象"小黑狗"是什么关系，对象"狗"与"狗尾巴"又是什么关系？

答：对象"狗"与对象"小黑狗"具有继承关系，即对象"小黑狗"继承了对象"狗"。"狗尾巴"是对象"狗"的一个属性，所以对象"狗"包含"狗尾巴"，二者是包含关系。

4. 简述 Java 语言的主要特点。

答：Java 语言的主要特点有：简单性、面向对象、分布式、健壮性、结构中立、安全性、可移植、解释的、高性能、多线程、动态性。

5. 简述 Java 语言与 C/C++语言的主要差异。

答：Java 基于 C++，与之有许多相似之处，但其设计更易于使用，它们之间的主要差异有：

（1）Java 无 C/C++中最复杂并有潜在危险的指针。

（2）Java 无 C/C++中的#include，#define 和头文件。

（3）Java 无 C/C++中的 structure,union 及 typedef。

（4）Java 无 C/C++中的函数、指针和多重继承。

（5）Java 无 C/C++中的 goto 指令。

（6）Java 无 C/C++中的操作符重载(operatior overloading)、自动类型的转换。

（7）Java 系统要求对对象进行相容性检查，以防止不安全的类型转换。

（8）Java 语言最强大的特性之一是它的平台独立性，Java 可以处理好平台之间的移植问题。

（9）Java 语言中没有全局变量的定义，只能通过公用的静态的变量实现，从而减少了引起错误的地方。

6. 什么叫 Java 虚拟机？什么叫 Java 字节码？

答：Java 虚拟机是一个软件系统，它可以翻译并运行 Java 字节码。它是 Java 的核心，保证了在任何异构的环境下都可运行 Java 程序，解决了 Java 的跨平台的问题。Java 的字节码(byte-code)是与平台无关的，是虚拟机的机器指令。

7. 简述 Java 程序的运行过程。

答：首先编写 Java 源代码程序，通过 Java 虚拟机编译成.class 的字节码程序。然后再执行翻译所生成的字节代码，属于先解释后执行的方式。在运行时，字节码载入器用于调入包含、继承所用到的所有类，确定内存分配，变成真正可执行的机器码。再通过字节码校验器检验字节码是否正确，如果通过校验，再由系统执行平台解释执行。

8. Java 程序分哪两类？各有什么特点？

答：Java 程序根据程序结构的组成和运行环境的不同可以分为两类：JavaApplication（Java 独立应用程序）和 JavaApplet（Java 小应用程序）。Java 独立应用程序是一个完整的程序，需要独立的 Java 解释器来解释执行；而 Java 小应用程序则是嵌在 Web 页面中的非独立应用程序，由 Web 浏览器内部所包含的 Java 解释器来解释执行，为 Web 页面增加交互性和动态性。

9. 根据自己的上机环境，简述 Java 程序的开发步骤。

答：Java 应用程序的运行经过编写、编译、运行三个步骤。

编写程序：使用记事本或其他软件编写程序的源代码，将源代码保存为 filename.java 文件。

编译程序：在 MS-DOS 命令窗口，将当前目录转换到 Java 源程序所在的保存目录；输入"javac filename.java"形式的命令进行程序编译。

执行程序：在同样的命令窗口中输入"javafilename"形式的命令执行程序。完成了程序的开发，查看相应目录，其中应该具有两个文件，分别是 XXXX.java 和 XXXX.class。

1.3 上 机 实 验

一、实验目的与意义

1. 掌握 JDK 平台的安装方法；
2. 掌握 Eclips 平台的安装方法；
3. 掌握在 Eclips 平台 Java 软件的开发方法。

二、实验内容

1. 安装 JDK 平台，并编译一个小程序；
2. 安装 Eclips 平台，并编译一个小程序。

三、实验要求

1. 从 Oracle 公司下载 Windows 版的 JDK7，并安装；
2. 从 Eclipse 网站下载 eclipse-jee-juno-SR1-win32.zip 并安装；
3. 利用 Eclips 开发平台，完成教材中的程序的编制与运行。

1.4 程序代码

一、安装 JDK

从 Oracle 公司下载 Windows 版的 JDK7，如图 1.1 所示。

http://www.oracle.com/technetwork/java/javase/downloads/jdk-7u3-download-1501626.html

（1）双击进入 JDK7 安装界面，点击下一步，如图 1.2 所示。

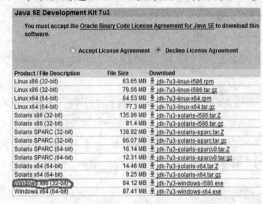

图 1.1　下载 JDK　　　　　　　　　图 1.2　JDK 安装向导界面

（2）点击"更改"按钮，根据用户实际情况选择安装路径，这里将路径名改为 C:\Program Files\Java\jdk1.7\。也可根据需要对开发工具、源代码、公共 JRE 作出相应修改，这里保持默认，如图 1.3 所示。点击"确定按钮"，开始安装 JDK，如图 1.4 所示。

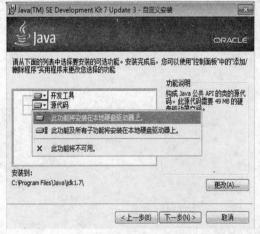

图 1.3　JDK 安装向导之更改文件夹　　　图 1.4　JDK 安装向导之自定义安装

（3）安装过程中会进入 jre7 安装路径选择界面，这里默认不变，选择"下一步"，如图 1.5 所示。

（4）安装结束，点击"继续"，如图 1.6 所示，进入 JavaFX SDK 设置界面，点击"取消"，如图 1.7 所示。

图 1.5　JDK 安装向导之选择安装路径

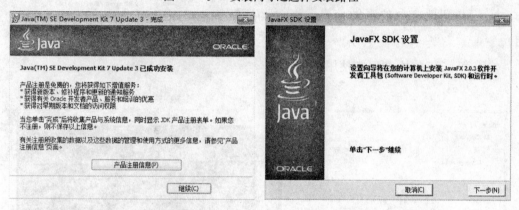

图 1.6　JDK 更新程序　　　　　图 1.7　JDK 安装完成

二、Java 运行环境变量设置

（1）点击计算机→右键"属性"→点击左侧"高级系统设置"→点击"环境变量"，如图 1.8 所示。

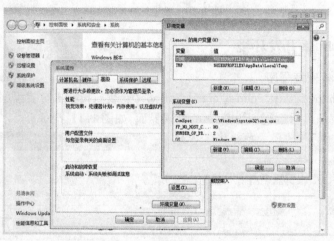

图 1.8　环境变量设置

（2）选择"系统变量"，点击"新建"，依次输入如下"变量名"、"变量值"：

JAVA_HOME：C:\Program Files\Java\jdk1.7（JDK 安装路径）

PATH：%JAVA_HOME%\bin；%JAVA_HOME%\jre\bin

CLASSPATH：.；%JAVA_HOME%\lib；%JAVA_HOME%\lib\tools.jar

点击确定,配置结束,如图1.9所示。

(3)验证结果。点击"开始",在搜索栏输入"cmd",回车进入命令行窗口,如图1.10所示。

在命令行窗口下键入以下命令,查看是否配置正确。

java-version:查看安装的JDK版本信息。

java:得到此命令的帮助信息。

javac:得到此命令的帮助信息。(一般需重启)

均正确,安装无误,可正常使用。

图1.9 编辑系统变量

图1.10 验证环境变量

三、安装 Eclipse 开发环境

1. 下载 Eclipse 安装程序

首先到网站 http://www.eclipse.org/downloads/下载 eclips 的安装程序 eclipse-jee-juno-SR1-win32.zip,解压缩后运行 eclipse.exe,出现界面,如图1.11界面。

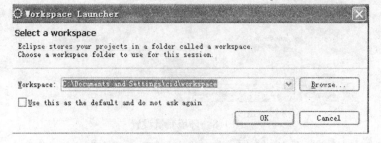

图1.11 Eclipse 的安装

这里是要我们设置工作目录,这个可以随便设置,这里我们采用默认配置,直接点击"OK"。

点击图1.12工作区右上角"Workbench"链接,进入工作台,如图1.13所示。

第 1 章　Java 语言概述

图 1.12　Eclips 工作区界面

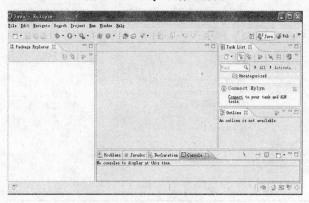

图 1.13　Eclipse 工作台

2. 现在我们来创建一个 Java 工程

　　File→new→javaproject（如果目录中没有"javaproject"，就选择"project"，然后在弹出的窗口中选择"java"→"javaproject"），如图 1.14 所示。

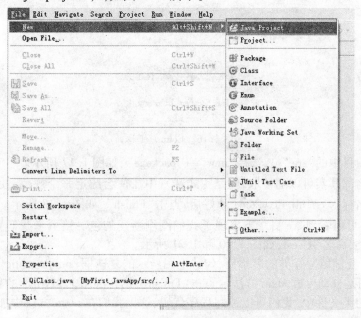

图 1.14　Eclipse 新建项目

图 1.15 中是 Eclipse 的新建工程配置界面,项目名称可以自己取,这里我们填:MyFirst_JavaApplication,其他的选项采用默认值,直接点击"Finish"。工作台,左边应该会有如图 1.16 所示的一个工程。

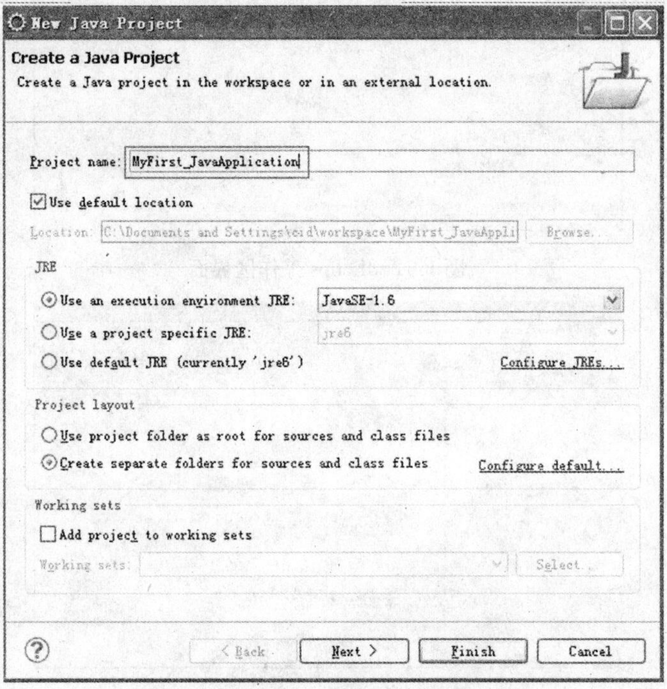

图 1.15 Eclipse 新建工程配置

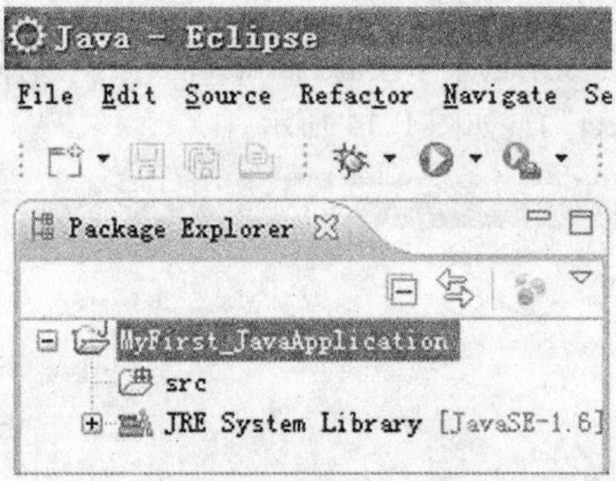

图 1.16 Eclipse 工程

右键单击项目中的"src"包,选择"new"→"package",如图 1.17 所示。

这里包的名称其实可以随便取,但是正式开发中往往会约定一个规范,便于项目的维护。这里我们的包名称为 cn.hlj.dfxy.tcr,填好后,点击"Finish",如图 1.18 所示。

如图 1.19,可以看到 src 目录下出现了我们刚刚新建的包。接下来我们要在刚刚创建的包中新建一个类,命名为 Test,如图 1.20 和 1.21 所示。

如图 1.21,我们填写好类的名称后,勾选上图所示的复选框(public static void main(String [ ] args)),目的是自动生成 main 方法。点击"Finish",出现如图 1.22 所示的界面。

在刚刚建的类文件中,我们可以编写程序代码,这里我们在 main 方法中加入如图 1.23 所示代码。

按"Ctrl + F11"运行程序。程序运行后,可以看到 Console 窗口中打印出了"Hello,

World!",如图 1.24 所示,至此第一个 Java 程序编写完成。

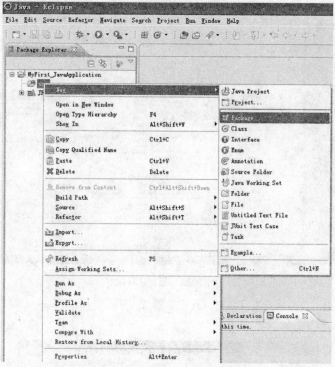

图 1.17 Eclipse 新建 Package

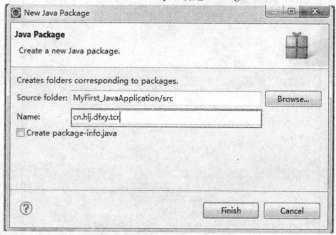

图 1.18 新建一个新的 Java 包

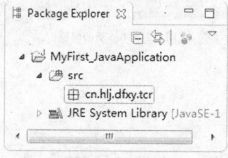

图 1.19 Java 新包

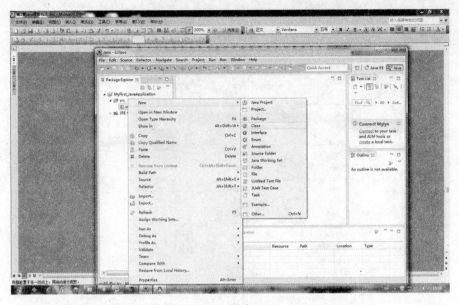

图 1.20 新建 Class

图 1.21 Test 类的配置

第1章 Java语言概述

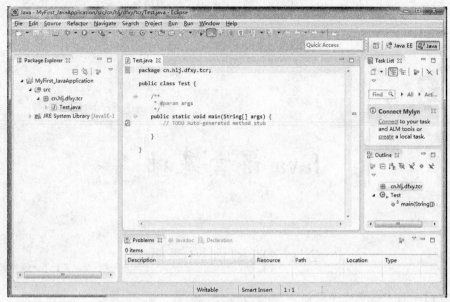

图 1.22 Test 类界面

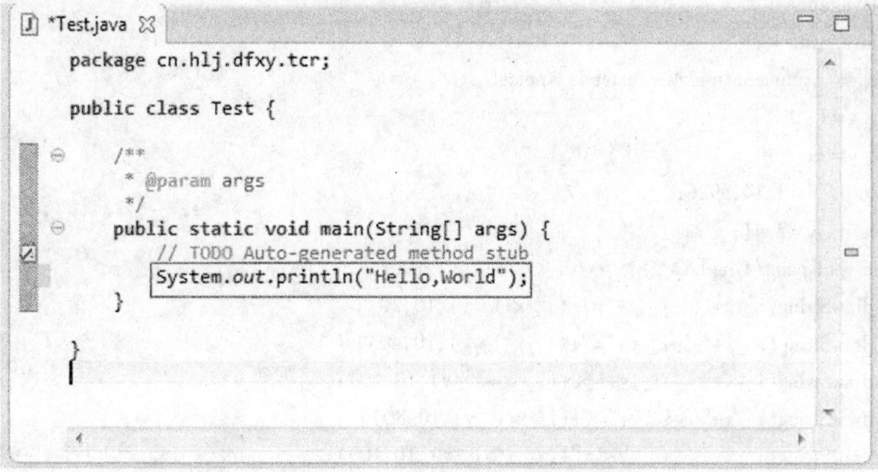

图 1.23 Test 类中编写程序

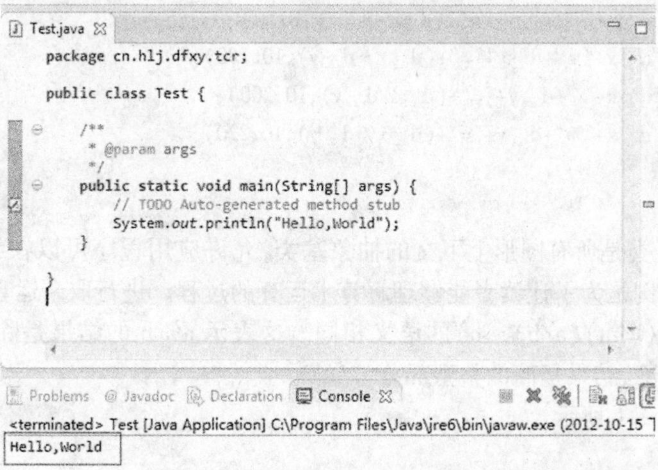

图 1.24 Eclipse 第一个程序运行

第 2 章 Java 语言基础

2.1 典型例题解析

【例 2.1】 算术运算程序示例

```java
import java.applet.*;
import java.awt.*;
public class ArithmeticOperation extends Applet{
    int i_x=1;
    int i_y=2;
    double d_x=3.1415926;
    double d_y=2.41;
    public void paint(Graphics g){
        g.drawString(i_x+"+"+i_y+"="+(i_x+i_y),10,20);
        g.drawString(i_x+"-"+i_y+"="+(i_x-i_y),10,40);
        g.drawString(i_x+"*"+i_y+"="+(i_x*i_y),10,60);
        g.drawString(i_x+"/"+i_y+"="+(i_x/i_y),10,80);
        g.drawString(i_x+"%"+i_y+"="+(i_x%i_y),10,100);
        g.drawString(d_x+"+"+d_y+"="+(d_x+d_y),10,140);
        g.drawString(d_x+"-"+d_y+"="+(d_x-d_y),10,160);
        g.drawString(d_x+"*"+d_y+"="+(d_x*d_y),10,180);
        g.drawString(d_x+"/"+d_y+"="+(d_x/d_y),10,200);
        g.drawString(d_x+"%"+d_y+"="+(d_x%d_y),10,220);
    }
}
```

解析：Graphics 类是所有图形上下文的抽象基类，允许应用程序可以在组件，以及闭屏图像上进行绘制。本例是为了让读者能够理解算术运算的过程。进行取余运算的操作数可以是浮点数，a%b 和 a-((int)(a/b)*b)的语义相同。这表示 a%b 的结果是除完后剩下的浮点数部分两个整型相除，结果是商的整型部分。

运行结果如图 2.1 所示。

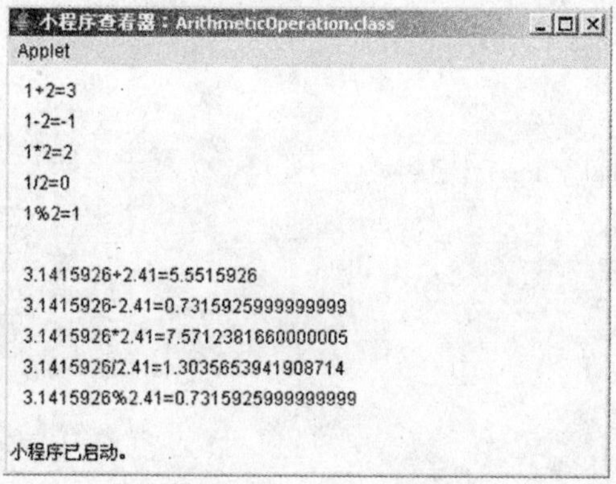

图 2.1 运行结果

【例 2.2】 编写一个具有 Java 不同数据类型和定义不同使用类型的变量、常量的类。

字符与逻辑类：

```
public class CharBooleanTest {
 public static void main(String[ ] args) {
   char c1 = 'C';
   System. out. println("c1 =" + c1);
   / * *
    * char 2 byte, ASCII 1 byte, ISO 8859-1 扩展 ASCII 2 byte
    * range 0 ～ 65536 (\u0000 ～ \u00FF) ISO 8859-1 字符
    */
   char c2 = '\u0041';
   System. out. println("c2 =" + c2);
   char c3 = 97;
   //char c3 = 324;
   boolean b = false;
   b = true;
   System. out. println(b);
 }
}
```

浮点类型类：

```
public class FloatTypeTest {
 public static void main(String[ ] args) {
   // TODO Auto-generated method stub
   float fa = 123.4f;
     //float fb = 12.5E300F;//--out of range
   float fb = (float)12.5E300;
   double da = 123; //int 自动转换为 double
   double db = 123.456D;
   double dc = 123.45e301;
   double dd = 243.343f; // float 自动转换为 double
   System. out. println(fa);
```

}

不同类型转换类：
```java
public class TypeCastTest {
  public static void main(String[] args) {
    char ca = 98;
    short cb = '我';
    short cc = 'a';
    short cd = '\u0012';
    short sa = (short)22.454;
    short sb = (short)324354;
    int ia = (int)34.3543;
    int ib = (int)1324343535432L;
    //以下转换可能会丢失精度
    float fa = 342322323;//int→float
    float fb = 2432432432421L;//long→float
    double da = 24324324323333343429L;//long→double
    double db = 2334343555664L;
    System.out.println(ca);
  }
}
```

【例2.3】 关系和逻辑运算示例程序
```java
import java.applet.*;
Import java.awt.*;
public class LogicalOperation extends Applet{
  boolean b1=true;
  boolean b2=false;
  int x1=3,y1=5;
  boolean b3=x1>y1&&x1++==y1--;
  public void paint(Graphics g){
    g.drawString(b1+"&&"+b2+"="+(b1&&b2),10,20);
    g.drawString(b1+"||"+b2+"="+(b1||b2),10,40);
    g.drawString("!"+b2+"="+(!b2),10,60);
    g.drawString("x="+x1+",y="+y1,10,80);
    g.drawString("(x>y&&x++==y--)="+b3+";x="+x1+",y="+y1,10,100);
  }
}
```

解析：关系运算符是比较两个数据大小关系的运算符，常用的关系运算符有>，>=，<，<=，==，!=。如果一个关系运算表达式，其运算结果是"真"，则表明该表达式所设定的大小关系成立；若运算结果为"假"，则说明了该表达式所设定的大小关系不成立。逻辑运算和关系运算的关系十分密切，关系运算是运算结果为布尔型量的运算，而逻辑运算是操作数和运算结果都是布尔型量的运算。

运行结果如图2.2所示。

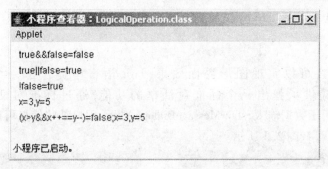

图 2.2　运行结果

【例 2.4】　一个标准输入输出的例子。一个简单字符输入输出程序 SimpleCharInOut.java

```
import java.io.*;
public class SimpleCharIntOut{
public static void main(String args[ ]){
    char c=' ';           //定义一个字符型变量初始化为空格
    System.out.print("Enter a character please:");//在屏幕上显示提示信息
    try{
        c=(char)System.in.read();//接受用户的键盘输入
    }catch(IOException e){
        System.err.println(e.toString());//可能抛出异常
    }
    System.out.println("You have entered character "+c);   //向用户屏幕输出字符
 }
}
```

解析:输入输出是程序的基本功能,与 C/C++相似,Java 语言中的输入输出涉及流的概念,借助流类实现输入输出。

System.in.read()能够接受用户的键盘输入,System.out.println()可以向屏幕输出字符。

运行结果如图 2.3 所示。

图 2.3　运行结果

【例 2.5】　将例题 2.4 用 JOptionPane 类进行改写,实现一个简单字符串输入输出程序 SimpleStringIntOut.java。

```
import javax.swing.JOptionPane;//导入 JOptionPane 类
public class SimpleStringIntOut{
    public static void main(String[ ] args) {
        String str;                            //定义一个字符串
        str=JOptionPane.showInputDialog("Input a String:");
        //从键盘输入一个字符串
        JOptionPane.showMessageDialog(null,"The String is:"+str,"Result is",
         JOptionPane.INFORMATION_MESSAGE);      //将字符串输出
```

```
        System.exit(0);                              //退出程序
    }
}
```

解析：Java 语言可以实现图形界面效果，Java 语言中提供了一个类 javax.swing.JOptionPane，该类提供了弹出一个标准对话框的功能，通过标准对话框来提示用户。showInputDialog( )用于数据输入；showMessageDialog( )提示用户某些信息可以由用户定义。

运行结果如图 2.4、图 2.5 所示。

图 2.4　例 2.5 的数据输入　　　　　图 2.5　例 2.5 的数据输出

2.2　课后习题解答

1. Java 语言对标识符命名有何规定？下面这些标识符哪些是合法的？哪些是不合法的？
（1）Myname1　（2）Java-Language　（3）2Person1　（4）_ is _ Has　（5）$12345

答：合法标识符：Myname1，_ is _ Has

非法标识符：Java-Language，2Person1，$12345

2. Java 有哪些基本数据类型？与 C/C++相比有何特点？复合数据类型有哪几种？

答：基本数据类型有：整型数据类型（字节整型、短整型、整型、长整型），实数数据类型（单精度实数、双精度实数），字符数据类型和布尔数据类型。与 C/C++相比，Java 的数据类型与 C++相似，但有两点不同：①在 Java 语言中所有的数据类型是确定的，与平台无关，所以在 Java 中无 sizeof 操作符；②Java 中每种数据类型都对应一个默认值。这两点体现了 Java 语言的跨平台性和完全稳定性。Java 的复合类型是由用户根据需要自己定义并实现其运算的数据类型，主要有类、接口和数组等。

3. Java 的注释方法有哪些？在实际开发中如何使用？

答：（1）单行（single-line）——短注释：//……

单独行注释：在代码中单起一行注释，注释前最好有一行空行，并与其后的代码具有一样的缩进层级。如果单行无法完成，则应采用块注释。

（2）块（block）注释：/*……*/

注释若干行，通常用于提供文件、方法、数据结构等的意义与用途的说明，或者算法的描述。

（3）文档注释：/**……*/

注释若干行，并写入 javadoc 文档。

4. Java 的字符类型采用何种编码方案？有何特点？

答：Java 的字符类型采用 16 位 Unicode（全球文字共享编码）方式，用 16 位来表示东西方字符。由于采用 Unicode 编码方案，使得 Java 在处理多语种的能力方面得到大大提高，从而为 Java 程序在基于不同语种之间实现平滑移植铺平了道路。

5. Java 有哪些运算符和表达式？请写出下面这些表达式的运算结果（设 a＝2，b＝3，f＝

true)。

(1)(--a)%b++ (2)(a>=1)&&a<=12? a:b) (3)f^(a>b) (4)(--a)<<a

答:Java 的运算符主要有算术运算符、关系运算符、条件运算符、位运算符、逻辑运算符以及赋值运算符。表达式是由运算符、操作数和方法调用,按照语言的语法规则构造而成的符号序列。表达式的结果是:

(1)--a%b++ 的结果是:1

(2)(a>=1)&&a<=12? a:b) 的结果是:1

(3) f^(a>b) 的结果是:false

(4)(--a)<<a 的结果是:0

6.阅读下列程序,写出运行结果。

```
public class Example2_1 {
    public static void main(String[] args){
        int i = 0;
        double x = 2.3;
        System.out.println("Result 1:"+(--i+i+i++));
        System.out.println("Result 2:"+(i+++i));
        i+=i+(i=4);
        System.out.println("Result 3:"+i);
        i=3+3*2%i--;
        System.out.println("Result 4:"+i);
        x+=1.2*3+x++;
        System.out.println("Result 5:"+x);
        x=x%3+4*2+x--;
        System.out.println("Result 6:"+x);
    }
}
```

运行结果:

i=0

Result 1:--i+i+i++=-1+-1+-1=-3(运算之后 i=0)

Result 2:i+++i=0+1=1(运算之后 i=1)

Result 3:i+=i+(i=4)等价于 i=i+(i+(i=4))这里 i 开始=1,然后把 i 赋值成4,所以 i=1+(1+4)=6(运算之后 i=6)

Result 4:i=3+3*2%i--=3+6%6=3+0=3(运算之后 i=3)

Result 5:x+=1.2*3+x++等价于 x=x+1.2*3+x++=2.3+3.6+2.3=8.2(运算之后 x=8.2,由于这是 double 类型的数据,所以里面 1.2*3 是个无限近似等于 3.6 的数)

Result 6:x=x % 3+4*2+x--=2.2+8+8.2=18.4

所以最后的结果是:

Result 1:-3

Result 2:1

Result 3:6

Result 4:3

Result 5:8.2

Result 6:18.4

2.3 上机实验

一、实验目的与意义

1. 了解 Java 的数据类型；
2. 掌握各种变量的声明方式；
3. 理解运算符的优先级；
4. 掌握 Java 基本数据类型、运算符与表达式、数组的使用方法；
5. 掌握 Java 注释的使用方法。

二、实验内容

1. 声明不同数据类型的变量。

程序功能：新建一个类 Demo2_1，定义 9 个变量，分别存放 0x55，0x55ff，1000000，0xffffL，'a'，0.23F，0.7E-3，true，"这是字符串类数据类型"等 9 个值，并将这些变量的值依次输出。

2. 使用运算符（关系、算术、赋值、位、逻辑）。

（1）新建一个类 Demo2_2，输入一个整数[0,9 999]，把该整数各位数字分别显示。

（2）新建一个类 Demo2_3，在打印 i 变量的同时分别对该变量进行 i++和 i--操作，同时每一步都应输出当前的 i 值。

（3）新建一个类 Demo2_4，给出一个数，写出此数左移、右移、无符号右移的表达式。

3. 使用表达式语句。

（1）i=3，j=4，分析表达式 20*8/4+i+j*i 的结果，并新建一个类 Demo2_5，编码测试。

（2）新建一个类 Demo2_6，根据给定的浮点类型的华氏温度值，用公式转化（摄氏=（华氏-32）*5/9）为摄氏温度。

三、实验要求

1. 编写一个声明 Java 不同数据类型变量的程序；
2. 编写一个使用运算符、表达式、变量的程序；
3. 编写一个使用 Java 数据的程序；
4. 编写表达式语句、复合语句的程序；
5. 为重要代码加上注释。

2.4 程序代码

1. 代码 Demo2_1

```
public class Demo2_1{
    public static void main(String[ ] args){
        byte b=0x55;
        short s=0x55ff;
        int i=1000000;
```

```java
        long l=0xffffL;
        char c='a';
        float f=0.23F;
        double d=0.7E-3;
        boolean B=true;
        String S="这是字符串类数据类型";
        System.out.println("字节型变量 b="+b);
        System.out.println("短整型变量 s="+s);
        System.out.println("整型变量 i="+i);
        System.out.println("长整型变量 l="+l);
        System.out.println("字符型变量 c="+c);
        System.out.println("浮点型变量 f="+f);
        System.out.println("双精度变量 d="+d);
        System.out.println("布尔型变量 B="+B);
        System.out.println("字符串类对象 S="+S);
    }
}
```

2. 代码 Demo2_2

```java
import java.io.IOException;
import java.util.Scanner;
public class Demo2_2{
    public static void main(String[] args) throws IOException{
        System.out.println("请输入一个0-9999 的数字:");
        Scanner reader=new Scanner(System.in);
        int a,d1,d2,d3,d4;//定义5个变量,a是输入的数据,其他分别是位数
        a=reader.nextInt();//从键盘获得数字
        d1=a/1000%10;    //求千位余数
        d2=a/100%10;     //求百位余数
        d3=a/10%10;      //求十位余数
        d4=a%10;         //求个位余数
        System.out.println("千位、百位、十位、个位分别是",a,d1,d2,d3,d4);
        //打印出结果
    }
}
```

3. 代码 Demo2_3

```java
package cn.hlj.dfxy.tcr;
public class Demo2_3{
    public static void main(String[] args){
        int i=1;//设置i的初值
        System.out.println("i:"+i);
        System.out.println(++i);//进行表达式加1,变量加1操作
        System.out.println(i++);//只进行变量加1操作,表达式不变
        System.out.println(--i);//进行表达式减1,变量加1操作
        System.out.println(i--);//只进行变量减1操作
```

    }
}

4. 代码 Demo 2_4

```java
public class Demo2_4 {
    public static void main(String[] args) {
        short a=0x34f2;
        int c;
        c=a>>3;    //右移三位
        System.out.println("a 右移 3 位为:"+c);
        c=a<<3;    //左移三位
        System.out.println("a 左移 3 位为:"+c);
        c=a>>>3;   //无符号右移三位
        System.out.println("a 无符号右移 3 位为:"+c);
    }
}
```

5. 代码 Demo2_5

```java
public class Demo2_5 {
    public static void main(String[] args) {
        int i=3;
        int j=4;
        System.out.println("表达式 20*8/4+i+j*i 的值为:"+20*8/4+i+j*i);
    }
}
```

6. 代码 Demo2_6

```java
import java.util.Scanner;
public class Demo2_6 {
    public static void main(String[] args) {
        float t1=0;
        System.out.println("请输入一个温度(华氏温度):");
        Scanner reader=new Scanner(System.in);
        t1=reader.nextFloat();
        System.out.println("华氏温度%d 的摄氏温度为:"+(t1-32)*5/9);
    }
}
```

第 3 章

基本控制结构

3.1 典型例题解析

一、使用选择语句

1. 使用 if...else 语句

(1)程序功能:使用 if...else 语句构造多分支,判断某一年是否为闰年。闰年的条件是符合下面二者之一:能被 4 整除,但不能被 100 整除;能被 4 整除,又能被 100 整除。

(2)编写源程序文件,代码如下:

```
public class Demo2_6 {
    public static void main( String args[ ] ) {
        boolean leap;
        int year = 2005;
        if (((year%4 = =0 && year%100! =0) || (year%400 = =0)))  //方法1
        System. out. println(year+"年是闰年");
        else
        System. out. println(year+"年不是闰年");
    }
}
```

(3)编译运行程序,其结果如图 3.1 所示。

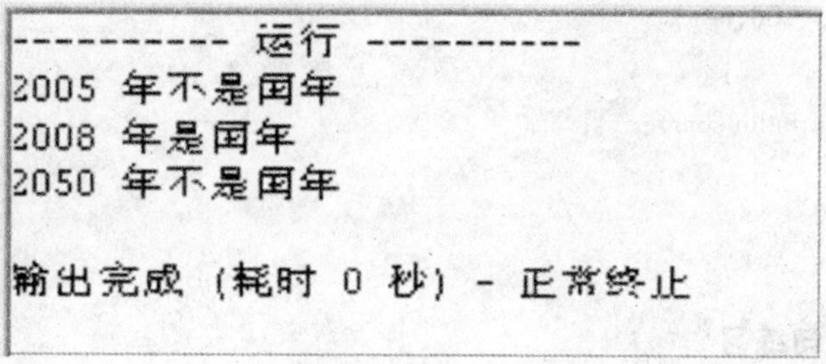

```
---------- 运行 ----------
2005 年不是闰年
2008 年是闰年
2050 年不是闰年

输出完成 (耗时 0 秒) - 正常终止
```

图 3.1 运行结果

2. 使用 switch 语句

(1)程序功能:在不同温度时显示不同的解释说明。

(2)程序源代码如下。
```java
class Demo2_7{
    public static void main(String args[]){
        int c=38;
        switch (c<10? 1:c<25? 2:c<35? 3:4) {
        case 1:
            System.out.println(" "+c+"℃ 有点冷。要多穿衣服。");
        case 2:
            System.out.println(" "+c+"℃ 正合适。出去玩吧。");
        case 3:
            System.out.println(" "+c+"℃ 有点热。");
        default:
            System.out.println(" "+c+"℃ 太热了！开空调。");
        }
    }
}
```

(3)编译运行程序,其结果如图3.2所示。

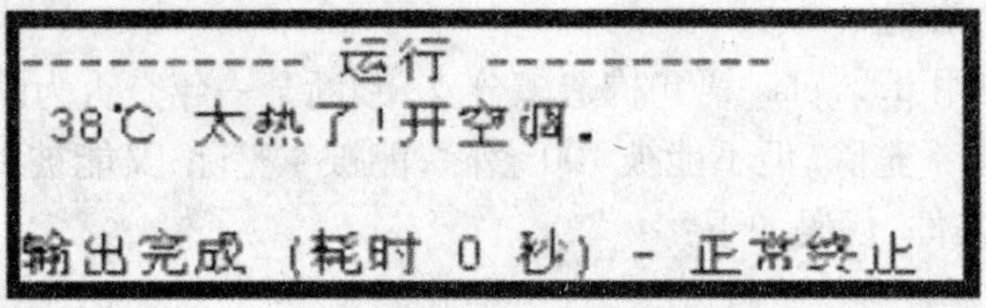

```
---------- 运行 ----------
38℃ 太热了!开空调.

输出完成 (耗时 0 秒) - 正常终止
```

图3.2 运行结果

二、使用循环语句

1. for 循环语句练习

(1)程序功能:将1至100之间所有的整数求和。
(2)程序源代码如下:
```java
public class Sum{
    public static void main(String []args){
        int sum=0;
        for(int i=1;i<=100;i++)
            sum=sum+i;
        System.out.println(sum);
    //i 不再有效
    }
}
```

2. while 循环语句练习

(1)程序功能:已知 sum(k)=1+2+3+…+k,问 k 为什么值时能够使得 sum(k)>2 000?
(2)程序源代码如下:
```java
public class FindMinimalK{
    public static void main(String []args){
```

```
        int sum=1;
        int k=1;
        while(sum<=2000){
            k++;
            sum=sum+k;
        }
        System.out.println("the minimal k="+k);
    }
}
```

3. do...while 循环语句练习

(1) 程序功能：求 1+2+…+100 之和，并将求和表达式与所求的和显示出来。

(2) 程序源代码如下：

```
class Demo2_10{
    public static void main(String args[]){
        int n=1,sum=0;
        do{
            sum+=n++;
        }while(n<=100);
        System.out.println("1+2+...+100="+sum);
    }
}
```

(3) 编译并运行程序，结果如图 3.3 所示。

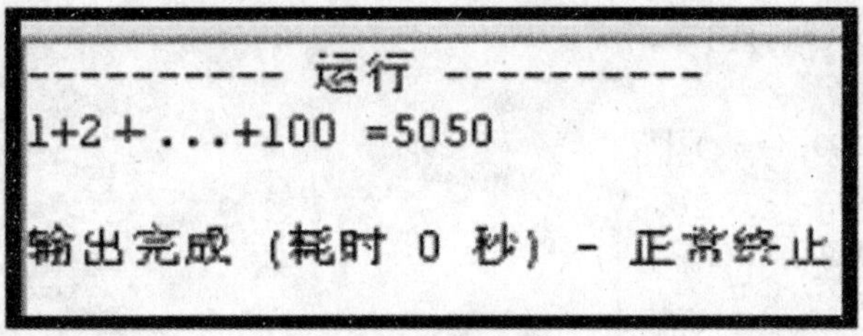

```
---------- 运行 ----------
1+2+...+100 =5050

输出完成 (耗时 0 秒) - 正常终止
```

图 3.3　运行结果

3.2　课后习题解答

一、选择题

1~6 DCBBDB

二、读程序，写结果

1. 女性　　未知
2. s=3367
3. s=105

三、编程题

1. 古典问题:有一对兔子,从出生后第 3 个月起每个月都生一对兔子,小兔子长到第 3 个月后每个月又生一对兔子,假如兔子都不死,问每个月的兔子总数为多少?

```java
//这是一个斐波那契数列问题
public class lianxi01 {
    public static void main(String[] args) {
        System.out.println("第1个月的兔子对数:    1");
        System.out.println("第2个月的兔子对数:    1");
        int f1 = 1, f2 = 1, f, M = 24;
        for(int i = 3; i <= M; i++) {
            f = f2;
            f2 = f1 + f2;
            f1 = f;
            System.out.println("第"+i+"个月的兔子对数:"+f2);
        }
    }
}
```

2. 判断 101~200 之间有多少个素数,并输出所有素数。

程序分析:判断素数的方法是用一个数分别去除 2 到 sqrt(这个数),如果能被整除,则表明此数不是素数,反之是素数。

```java
public class lianxi02 {
    public static void main(String[] args) {
        int count = 0;
        for(int i = 101; i < 200; i += 2) {
            boolean b = false;
            for(int j = 2; j <= Math.sqrt(i); j++)
            {
                if(i % j == 0) { b = false; break; }
                else           { b = true; }
                if(b == true) {count++; System.out.println(i);}
            }
        }
        System.out.println("素数个数是:"+count);
    }
}
```

3. 打印出所有的"水仙花数",所谓"水仙花数"是指一个三位数,其各位数字立方和等于该数本身。例如,153 是一个"水仙花数",因为 $153 = 1^3 + 3^3 + 5^3$。

```java
public class lianxi03 {
    public static void main(String[] args) {
        int b1, b2, b3;
        for(int m = 101; m < 1000; m++) {
            b3 = m / 100;
            b2 = m % 100 / 10;
```

```
      b1 = m % 10;
      if((b3 * b3 * b3+b2 * b2 * b2+b1 * b1 * b1)= =m) {
      System. out. println(m+"是一个水仙花数");
      }
     }
    }
   }
```

4. 将一个正整数分解质因数。例如,输入 90,打印出 90 = 2 * 3 * 3 * 5。

程序分析:对 n 进行分解质因数,应先找到一个最小的质数 k,然后按下述步骤完成:

(1)如果这个质数恰等于 n,则说明分解质因数的过程已经结束,打印出即可。

(2)如果 n <> k,但 n 能被 k 整除,则应打印出 k 的值,并用 n 除以 k 的商,作为新的正整数 n,重复执行第一步。

(3)如果 n 不能被 k 整除,则用 k+1 作为 k 的值,重复执行第一步。

```
import java. util. * ;
public    class    lianxi04 {
  public static void main(String[ ] args) {
    Scanner s = new Scanner(System. in);
    System. out. print( "请键入一个正整数: ");
    int     n = s. nextInt( );
    int k = 2;
    System. out. print(n+"=");
    while(k <=n) {
      if(k= =n) {System. out. println(n);break;}
      else if( n % k= =0) {System. out. print(k+" * ");n = n / k; }
      else    k++;
    }
  }
}
```

5. 利用条件运算符的嵌套来完成此题:学习成绩>=90 分的同学用 A 表示,60~89 分之间的用 B 表示,60 分以下的用 C 表示。

```
import java. util. * ;
public class lianxi05 {
public static void main(String[ ] args) {
    int x;
    char grade;
    Scanner s = new Scanner(System. in);
    System. out. print( "请输入一个成绩:");
    x = s. nextInt( );
    grade = x >=90 ? 'A' : x >=60 ? 'B':'C';
    System. out. println("等级为:"+grade);
  }
}
```

6. 输入两个正整数 m 和 n,求其最大公约数和最小公倍数。

/*在循环中,只要除数不等于0,用较大的数除以较小的数,将小的一个数作为下一轮循环的大数,取得的余数作为下一轮循环的较小的数,如此循环直到较小的数的值为0,返回较大的数,此数即为最大公约数,最小公倍数为两数之积除以最大公约数。*/

```java
import java.util.*;
public class lianxi06 {
    public static void main(String[] args) {
        int a,b,m;
        Scanner s=new Scanner(System.in);
        System.out.print("键入一个整数:");
        a=s.nextInt();
        System.out.print("再键入一个整数:");
        b=s.nextInt();
        deff cd=new deff();
        m=cd.deff(a,b);
        int n=a*b/m;
        System.out.println("最大公约数:"+m);
        System.out.println("最小公倍数:"+n);
    }
}
class deff {
    public int deff(int x, int y) {
        int t;
        if(x < y) {
            t=x;
            x=y;
            y=t;
        }
        while(y!=0) {
            if(x==y) return x;
            else {
                int k=x%y;
                x=y;
                y=k;
            }
        }
        return x;
    }
}
```

7. 输入一行字符,分别统计出其中英文字母、空格、数字和其他字符的个数。

```java
import java.util.*;
public class lianxi07 {
    public static void main(String[] args) {
        int digital=0;
```

```
    int character=0;
    int other=0;
    int blank=0;
    char[ ] ch=null;
    Scanner sc=new Scanner(System.in);
    String s=sc.nextLine();
    ch=s.toCharArray();
    for(int i=0; i<ch.length; i++){
      if(ch >='0' && ch <='9'){
        digital++;
      }
      else if((ch >='a' && ch <='z') || ch > 'A' && ch <='Z'){
        character++;
      } else if(ch==' '){
        blank++;
      } else {
        other++;
      }
    }
    System.out.println("数字个数："+digital);
    System.out.println("英文字母个数："+character);
    System.out.println("空格个数："+blank);
    System.out.println("其他字符个数:"+other);
  }
}
```

8. 求 s=a+aa+aaa+aaaa+aa...a 的值,其中 a 是一个数字。例如,2+22+222+2 222+22 222（此时共有 5 个数相加）,几个数相加由键盘控制。

```
import java.util.*;
public class lianxi08{
  public static void main(String[] args){
    long a, b=0, sum=0;
    Scanner s=new Scanner(System.in);
    System.out.print("输入数字 a 的值：");
    a=s.nextInt();
    System.out.print("输入相加的项数:");
    int n=s.nextInt();
    int i=0;
    while(i < n){
      b=b+a;
      sum=sum+b;
      a=a * 10;
      ++i;
    }
    System.out.println(sum);
```

```
    }
}
```

9. 一个数如果恰好等于它的因子之和，这个数就称为"完数"。例如，6=1+2+3。编程找出 1 000 以内的所有完数。

```java
public class lianxi09 {
    public static void main(String[] args) {
        System.out.println("1 到 1000 的完数有：");
        for(int i=1; i<1000; i++) {
            int t=0;
            for(int j=1; j<=i/2; j++) {
                if(i % j == 0) {
                    t=t+j;
                }
            }
            if(t==i) {
                System.out.print(i+"    ");
            }
        }
    }
}
```

10. 一球从 100 m 高度自由落下，每次落地后反跳回原高度的一半；再落下，求它在第 10 次落地时，共经过多少米？第 10 次反弹多高？

```java
public class lianxi10 {
    public static void main(String[] args) {
        double h=100, s=100;
        for(int i=1; i<10; i++) {
            s=s+h;
            h=h/2;
        }
        System.out.println("经过路程:"+s);
        System.out.println("反弹高度:"+h/2);
    }
}
```

3.3 上机实验

一、实验目的与意义

1. 理解 Java 程序语法结构；
2. 掌握顺序结构、选择结构和循环结构语法的程序设计方法。

二、实验内容

1. 输出 9*9 口诀。

2. 猴子吃桃问题:猴子第一天摘下若干个桃子,当即吃了一半,还不过瘾,又多吃了一个,第二天早上又将剩下的桃子吃掉一半,又多吃了一个。以后每天早上都吃了前一天剩下的一半零一个。到第 10 天早上想再吃时,见只剩下一个桃子了。求第一天共摘了多少桃子?

3. 输入某年某月某日,判断这一天是这一年的第几天?

4. 输入三个整数 x,y,z,请把这三个数由小到大输出。

5. 一个 5 位数,判断它是不是回文数(12321 即是回文数,个位与万位相同,十位与千位相同)。

三、实验要求

1. JDK1.5 与 eclipse 开发工具;

2. 编写使用不同选择结构的程序,编写使用不同循环结构的程序。

3.4 程序代码

1. 输出 9*9 口诀。

```java
public class lianxi1{
    public static void main(String[] args){
        for(int i=1; i<10; i++){
            for(int j=1; j<=i; j++){
                System.out.print(j+"*"+i+"="+j*i+"    ");
                if(j*i<10){System.out.print(" ");}
            }
            System.out.println();
        }
    }
}
```

2. 猴子吃桃问题:猴子第一天摘下若干个桃子,当即吃了一半,还不过瘾,又多吃了一个,第二天早上又将剩下的桃子吃掉一半,又多吃了一个。以后每天早上都吃了前一天剩下的一半零一个。到第 10 天早上想再吃时,见只剩下一个桃子了。求第一天共摘了多少桃子?

```java
public class lianxi2{
    public static void main(String[] args){
        int x=1;
        for(int i=2; i<=10; i++){
            x=(x+1)*2;
        }
        System.out.println("猴子第一天摘了"+x+"个桃子");
    }
}
```

3. 输入某年某月某日,判断这一天是这一年的第几天?

```java
import java.util.*;
public class lianxi14{
    public static void main(String[] args){
```

```java
int year, month, day;
int days=0;
int d=0;
int e;
input fymd=new input( );
do{
    e=0;
    System.out.print("输入年:");
    year=fymd.input( );
    System.out.print("输入月:");
    month=fymd.input( );
    System.out.print("输入天:");
    day=fymd.input( );
    if (year < 0 || month < 0 || month > 12 || day < 0 || day > 31){
        System.out.println("输入错误,请重新输入!");
        e=1 ;
    }
}while( e==1);
for (int i=1; i<month; i++){
    switch (i){
        case 1:
        case 3:
        case 5:
        case 7:
        case 8:
        case 10:
        case 12:
            days=31;
            break;
        case 4:
        case 6:
        case 9:
        case 11:
            days=30;
            break;
        case 2:
            if (((year % 400==0) || (year % 4==0 && year % 100 !=0))){
                days=29;
            }else{
                days=28;
            }
            break;
    }
    d+=days;
```

```
        }
System. out. println(year+"-"+month+"-"+day+"是这年的第"+(d+day)+"天。");
    }
}
class input{
    public int input( ) {
        int value=0;
        Scanner s=new Scanner(System. in);
        value=s. nextInt( );
        return value;
    }
}
```

4. 输入三个整数 x,y,z,请把这三个数由小到大输出。

```
import java. util. * ;
public class lianxi15 {
    public static void main(String[ ] args) {
        input fnc=new input( );
        int x=0, y=0, z=0;
        System. out. print("输入第一个数字:");
        x=fnc. input( );
        System. out. print("输入第二个数字:");
        y=fnc. input( );
        System. out. print("输入第三个数字:");
        z=fnc. input( );
        if(x > y) {
          int t=x;
          x=y;
          y=t;
        }
        if(x > z) {
          int t=x;
          x=z;
          z=t;
        }
        if(y > z) {
          int t=y;
          y=z;
          z=t;
        }
        System. out. println( "三个数字由小到大排列为:"+x+" "+y+" "+z);
    }
}
    class input{
        public int input( ) {
```

```
            int value=0;
            Scanner s=new Scanner(System.in);
            value=s.nextInt();
            return value;
        }
    }
```

5. 一个5位数,判断它是不是回文数(12321 即是回文数,个位与万位相同,十位与千位相同)。

```
import java.util.*;
public class lianxi25 {
    public static void main(String[] args) {
        Scanner s=new Scanner(System.in);
        int a;
        do{
            System.out.print("请输入一个5位正整数:");
            a=s.nextInt();
        }while(a<10000||a>99999);
        String ss=String.valueOf(a);
        char[] ch=ss.toCharArray();
        if(ch[0]==ch[4]&&ch[1]==ch[3]){
            System.out.println("这是一个回文数");}
        else {System.out.println("这不是一个回文数");}
    }
}
//这个更好,不限位数
import java.util.*;
public class lianxi25a {
    public static void main(String[] args) {
        Scanner s=new Scanner(System.in);
        boolean is=true;
        System.out.print("请输入一个正整数:");
        long a=s.nextLong();
        String ss=Long.toString(a);
        char[] ch=ss.toCharArray();
        int j=ch.length;
        for(int i=0; i<j/2; i++) {
            if(ch[i]!=ch[j-i-1]){is=false;}
        }
        if(is==true){System.out.println("这是一个回文数");}
        else {System.out.println("这不是一个回文数");}
    }
}
```

第4章 数组、方法与字符串

4.1 典型例题解析

【例4.1】 编程使用字符数组和字符串分别输出字符串"ABCDE"。

解析:字符串可以直接输出,也可以通过方法 charAt( ) 逐个字符输出。

```
//CharArray. java
public class CharArray {
  public static void main(String[ ] args) {
    String s = new String("ABCDE");
    char[ ] a;
    a = s.toCharArray( );
    System. out. println("s="+s);
    System. out. println("s. length( ) = "+s. length( )+"    a. length = "+a. length);
    for ( int i = 0; i < s. length( ); i++) {
      System. out. println("s. charAt("+i+") = "+s. charAt(i)+"    a["+i
        +"] = "+a[i]);
    }
  }
}
```

运行结果:
```
s = ABCDE
s. length( ) = 5    a. length = 5
s. charAt(0) = A    a[0] = A
s. charAt(1) = B    a[1] = B
s. charAt(2) = C    a[2] = C
s. charAt(3) = D    a[3] = D
s. charAt(4) = E    a[4] = E
```

【例4.2】 随机产生1到6的整数,运行1 000次,模拟骰子6面出现的概率。

解析:在这个例子中,利用 java. util 包中的 API 类 Random 来产生随机数。Random 提供了 nextInt(n) 产生从 0 到 n-1 的随机数。

```
import java. util. Random;
```

```java
public class DieStatisticsTest {
    public static void main(String[] args) {
        int side = 1;
        int[] frequencies = new int[6];
        Random randomNumber = new Random();
        for (int roll = 1; roll <= 10000; roll++)
            ++frequencies[randomNumber.nextInt(6)];
        System.out.println("Side\t"+"Frequency");
        for (int frequency : frequencies)
            System.out.println(side+++"\t"+frequency);
    }
}
```

运行结果：
Side Frequency
1 1615
2 1686
3 1632
4 1716
5 1658
6 1693

【例4.3】 建立一个3×4的矩阵，并查找值最大的矩阵元素。

解析：要存放矩阵，需要使用二维数组。本例中，定义了一个3行4列的二维数组 matrix 来存放矩阵。通过遍历二维数组，找到矩阵中的最大元素值。

在create()方法中，使用双重循环产生12个0～200之间的随机整数，并放入二维数组 matrix 的各元素中。外层循环控制行，内层循环控制列。

在output()方法中，使用双重循环把矩阵元素输出。外层循环控制行，内层循环控制列。

在定义 max()方法中，使用双重循环遍历二维数组查找最大元素值一。外层循环控制行，内层循环控制列。

```java
public class Matrix {
    public static void create(int[][] matrix) {
        for(int row = 0; row < matrix.length; row++) {
            int column = 0;
            while(column < matrix[row].length) {
                matrix[row][column] = (int)(Math.random()*200);
                column = column+1;
            }
        }
    }
    public static void output(int[][] matrix) {
        System.out.println("矩阵：");
        for(int row = 0; row < matrix.length; row++)
            for(int column = 0; column < matrix[row].length; column++)
                System.out.print(matrix[row][column]+"\t");
```

```
            System.out.println( );
        }
    }
    public static int max(int[ ][ ] matrix){
        int result=matrix[0][0];
        for(int row=0;row<matrix.length;row++)
            for(int column=0;column<matrix[row].length;column++)
            if(matrix[row][column]>result)
                result=matrix[row][column];
            return result;
    }
    public static void main(String[ ] args){
        int[ ][ ] matrix=new int[3][4];
        create(matrix);
        output(matrix);
        System.out.println("最大值为:"+max(matrix));
    }
}
```

运行结果:
矩阵:
17   93   16   134
51   88   121 195
23   162 26   127
最大值为:195

【例 4.4】 判断数组元素是否重复

要求:判断一个数组中是否存在相同的元素,如果存在相同的元素则输出"重复",否则输出"不重复"。

解析:该题中如果需要判断数组中元素是否重复,则需要对数组中的元素进行两两比较,如果有任意一组元素相等,则该数组中的元素存在重复,如果任意一组元素都不相等,则表示数组中的元素不重复。

假设数组中的元素不重复,两两比较数组中的元素,使用数组中的第一个元素和后续所有元素比较,接着使用数组中的第二个元素和后续元素比较,以此类推实现两两比较,如果有一组元素相同,则数组中存储重复,结束循环。把比较的结果存储在一个标志变量里,最后判断标志变量的值即可。

```
public class Matrix{
public class circle{
public static void main(String[ ] args){
    int[ ] n={1,2,3,1,0};
    boolean flag=true;  //假设不重复
    for(int i=0;i < n.length-1;i++){  //循环开始元素
        for(int j=i+1;j < n.length;j++){  //循环后续所有元素
        //如果相等,则重复
        if(n[i]==n[j]){
```

```
                flag=false;//设置标志变量为重复
                break;//结束循环
            }
        }
    }
    //判断标志变量
    if(flag){
        System.out.println("不重复");
    }else{
        System.out.println("重复");
    }
}
```

运行结果:
重复

【例4.5】 判断数组是否对称

解析:该题中用于判断数组中的元素关于中心对称,也就是说数组中的第一个元素和最后一个元素相同,数组中的第二个元素和倒数第二个元素相同,以此类推,如果比较到中间,所有的元素都相同,则数组对称。实现思路:把数组长度的一半作为循环的次数,假设变量 i 从 0 循环到数组的中心,则对应元素的下标就是数组长度-i-1,如果对应的元素有一组不相等则数组不对称,如果所有对应元素都相同,则对称。

```
public class Matrix{
    public class double{
        int[] n={1,2,0,2,1};
        boolean flag=true;//假设对称
        for(int i=0;i< n.length/2;i++){//循环数组长度的一半
            //比较元素
            if(n[i]! =n[n.length - i - 1]){
                flag=false;//不对称
                break;//结束循环
            }
        }
        if(flag){
            System.out.println("对称");
        }else{
            System.out.println("不对称");
        }
    }
}
```

运行结果:
对称

【例4.6】 邮票组合。某人有4张3分和3张5分的邮票。编写一个Java程序,计算用这些邮票可以得到多少种不同的邮资?

解析:对此问题可以建立如下数学模型:

$s = 3t + 5f$

其中,t=0,1,2,3,4 为 3 分邮票张数,f=0,1,2,3 为 5 分邮票张数。这是一个穷举问题。

```java
public class Stamp {
  public static void printKinds() {
    int [ ]kinds=new int[20];
    int t,f,v=0;
    int i,n=0;
    for(t=0;t<=4;t++)
      for(f=0;f<=3;f++){
        v=t*3+f*5;
        for(i=0;kinds[i]!=0;i++)
          if(v==kinds[i])
          break;
        if(kinds[i]==0&&v!=0){
          kinds[i]=v;
          n++;
        }
      }
    System.out.println("方案数="+n);
    for(i=0;kinds[i]!=0;i++)
      System.out.println("方案"+i+": "+kinds[i]);
  }
  public static void main(String args[ ]){
    Stamp s=new Stamp();
    s.printKinds();
  }
}
```

运行结果:

方案数=19
方案 0: 5
方案 1: 10
方案 2: 15
方案 3: 3
方案 4: 8
方案 5: 13
方案 6: 18
方案 7: 6
方案 8: 11
方案 9: 16
方案 10: 21
方案 11: 9

方案 12: 14
方案 13: 19
方案 14: 24
方案 15: 12
方案 16: 17
方案 17: 22
方案 18: 27

4.2 课后习题解答

一、填空题

1. 下标　数组名　数据类型
2. 24
3. 二维　3　2　5

二、编程题

1. 编程对 10 个整数进行排序。

解析:静态方法 bubbleSort()对一维数组进行由小到大的排序,使用了冒泡排序算法。其基本思想是:将相邻的两个元素进行比较,若次序不对,则将两个元素的值互相交换。main()方法调用 bubbleSort()方法对数组 m 进行排序,并输出了排序前后的结果供比较。

```
public static ovid bubbleSort(int[ ] a){
  int n=a.length;
  int temp;
  for(int i=n-1;i>0;i--)
  for(int j=0;j<i-1;j++){
    if(a[j+1]<a[i])
    {
      temp=a[j+1];
      a[j+1]=a[j];
      a[j]=temp;
    }
  }
}
public static void main(String[ ] args)
{
  int[ ] m={10,8,21,65,23,52,78,83,30,99};
  System.out.println("排序前的数组是:");
  for(int i=0;i<m.length;i++){
    System.out.print(m[i]+"   ");
    System.out.println();
    bubbleSort(m);
    System.out.println("排序后的数组是:\n");
```

```
        for(int i=0;i<m.length;i++)
            System.out.print(m[i]+"   ");
        }
    }
}
```
运行结果:
排序前的数组是:
10 8 21 65 23 52 78 83 30 99
排序后的数组是:
8 10 21 23 30 52 65 78 83 99

2.下面哪些语句是合法的数组声明?

(1) int i = new int(30);
(2) double d[ ] = new double[30];
(3) int i[ ] = (3,4,3,2);
(4) float f[ ] = {2.3,4.5,5.6};
(5) char[ ] c = new char( );

答案:(1)非法　　　　　　　int i = new int[30];
　　　(2)合法
　　　(3)非法　　　　　　　int i[ ] = {3,4,3,2};
　　　(4)合法
　　　(5)非法　　　　　　　char[ ] c = new char[10];

3.求一个10行、10列整型方阵对角线上元素之积。

解析:程序首先创建一个10行10列的数组a,利用随机数方法Math.random()对数组的各元素赋值。方阵主对角线上元素的下标相同,副对角线上行列下标和为9(因为数组元素下标从0开始),利用这两个特性对方阵主对角线和副对角线上的元素进行累积,最后输出结果。

```
public class MatrixCalculation{
    public static void main(String[ ] args)
    {
        int a[ ][ ] = new int[10][10];
        System.out.println("随机生成的方阵为\n");
        for(int i=0;i<10;i++)
        {
            for(int j=0;j<10;j++){
                a[i][j] = (int)(10 * Math.random())+1;
                System.out.print(a[i][j]+"   ");
            }
            System.out.println("\n");
        }
        int sum = 1;
        for(int i=0;i<10;i++)
            sum *= a[i][j];
```

```
        System.out.println("主对角线元素的元素各为:"+sum);
        int num=1;
        for(int i=0;i<10;i++)
            num*=a[9-i][i];
        System.out.println("副对角线上元素积为:"+num);
    }
}
```

运行结果：
随机生成的方阵为
3 7 3 7 10 8 2 2 8 3
8 3 8 3 10 9 4 9 2 2
5 9 6 8 10 8 9 1 9 1
1 1 9 5

主对角线上元素积为:2624400
副对角线上元素积为:108864

4. 实现矩阵转置，即将矩阵的行、列互换，一个 m 行 n 列的矩阵将转换为 n 行 m 列。

解析：程序通过命令行读入矩阵的行数 m 和列数 n，构造一个 m 行 n 列的矩阵 a 和一个 n 行 m 列的矩阵 b，利用 Math.random() 对矩阵 a 的各元素赋随机值，对矩阵 a 进行转置，然后输出转置后的矩阵 b。

```
public class Transpose
{
    public static ovid main(String[] args)
    {
        int m,n;
        if(args.length!=2)
        {
            System.out.println("输入格式错误！请按照此格式输入:javaTransposs m n");
            System.exit(0);
        }
        m=Integer.parseInt(args[0]);
        n=Integer.parseInt(args[1]);
        int a[][]=new int[m][n];
        int b[][]=new int[n][m];
        System.out.println("转置前的矩阵\n");
        for(int i=0;i<m;i++)
        {for(int j=0;j<n;j++)
            {
                a[i][j]=(int)(100*Math.random());
                System.out.print(a[i][j]+"  ");
            }
            System.out.println(" \n");
        }
        for(int i=0;i<n;i++)
```

```
        }
        for(int j=0;j<m;j++)
            System.out.print(b[i][j]+"  ");
            System.out.println("\n");
    }
  }
}
```

运行结果：
转置前的矩阵
36　98　44　70
70　38　66　63
87　25　3　65
转置后的矩阵
36　70　87
98　38　25
44　66　3
70　63　65

5. 编写程序，计算一维数组中的最大值与其所在的位置。

解析：程序首先通过命令行读入一个整数 m，构造一个有 m 个元素的一维数组 a。使用随机数方法 Math.random( )对数组 a 各元素赋值。定义变量 max 和 sit 分别存放最大值和最大值的位置。对数组元素从 a[1]开始，逐个与 max 进行比较。如果比 max 大，则将此元素值赋给 max，并把位置赋给 sit，循环比较直到 a[m-1]结束。最后输出 max 以及它的位置。

```
public class FindMax
{
    public static void main(String[] args)
    {
        if(args.length!=1)
        {
            System.out.println("输入格式错误！请按照此格式输入:javaFindMax m");
            System.exit(0);
        }
        int m=Integer.parseInt(args[0]);
        int a[]=new int[m];
        System.out.println("随机生成的一维数组:");
        for(int i=0;i<m;i++)
        {
            a[i]=(int)(100*Math.random());
            System.out.print(a[i]+"  ");
        }
        int max,sit;
        max=a[0];
        sit=0;
        for(int i=1;i<m;i++)
```

```
            if(a[i]>=max)
            {
                max=a[i];
                sit=i;
            }
        }
        System.out.println("\n 此数组的最大值是:"+max+"位置是:"+sit);
    }
}
```

运行结果：

随机生成的一维数组

89 36 6 43 14 20

此数组中最大值是:89 位置是:0。

6. 从键盘上输入 10 个双精度浮点数后,求出这 10 个数的和以及它们的平均值。要求分别编写相应求和及求平均值的方法。

```
import java.io.*;
public class Test
{
    public static void main(String[] args) throws IOException
    {
        InputStreamReader reader=new InputStreamReader(System.in);
        BufferedReader input=new BufferedReader(reader);
        String temp;
        double x[]=new double[10];
        for(int i=0;i<10;i++)
        {
            temp=input.readLine();
            x[i]=Double.parseDouble(temp);
        }
        S(x);
        Avg(x);
    }
    public static void S(double x[])
    {
        //计算并输出和
        double sum=0;
        for(int i=0;i<10;i++)
        {
            sum+=x[i];
        }
        System.out.println("总和:"+sum);
    }
```

```java
    public static void Avg(double x[])
    {
        //计算并输出平均数
        double sum=0;
        for(int i=0;i<10;i++)
        {
            sum+=x[i];
        }
        System.out.println("平均数:"+sum/10);
    }
}
```

7. 编写一个方法,实现将字符数组倒序排列,即进行反序存放。

```java
import java.io.*;
public class Test
{
    public static void main(String[] args) throws IOException
    {
        char c[]={'O','l','y','m','p','i','c',' ','G','a','m','e','s'};
        rever(c);
        System.out.println(c);
    }
    public static void rever(char c[])
    {
        char t;
        for(int i=0,j=c.length-1;i<j;i++,j--)
        {
            t=c[i];
            c[i]=c[j];
            c[j]=t;
        }
    }
}
```

三、简答题

1. Java 语言为什么要引入方法这种编程结构?

答:提高复用度,减少程序代码量,促进程序结构化,提高可读性和可维护性。

2. 为什么要引入数组结构,数组有哪些特点? Java 语言创建数组的方式有哪些?

答:为了便于处理一批同类型的数据,Java 语言引入了数组类型:

首先,数组中的每个元素都是相同数据类型的;其次,数组中的这些相同数据类型元素是通过数组下标来标识的,并且该下标是从 0 开始的;最后,数组元素在内存中的存放是连续的。

Java 语言规定,创建数组可以有两种方式:初始化方式和 new 操作符方式。初始化方式是指直接给数组的每一个元素指定一个初始值,系统自动根据所给出的数据个数为数组分配相应的存储空间,通常这样创建数组的方式适用于数组元素较少的情形。对于数组比较大的情形,即数组元素过多,用初始化方式显然不妥,这时应采用第二种方式,即 new 操作符方式。

四、写出程序的运行结果

1. 有如下四个字符串 s1,s2,s3 和 s4:
```
String s1 = "Hello World! ";
String s2 = new String("Hello World! ");
s3 = s1;
s4 = s2;
```
求下列表达式的结果是什么?
```
s1 = = s3
s3 = = s4
s1 = = s2
s1. equals(s2)
s1. compareTo(s2)
```
答案:表达式的结果是:
false
false
false
true
0

2. 下面程序输出的结果是什么?
```
public class Test {
    public static void main(String[ ] args) {
        String s1 = "I like cat";
        StringBuffer sb1 = new StringBuffer ("It is Java");
        String s2;
        StringBuffer sb2;
        s2 = s1. replaceAll("cat","dog");
        sb2 = sb1. delete(2,4);
        System. out. println("s1 为:"+s1);
        System. out. println("s2 为:"+s2);
        System. out. println("sb1 为:"+s1);
        System. out. println("sb2 为:"+s2);
    }
}
```
答案:程序的输出结果为:
s1 为:I like cat
s2 为:I like dog
sb1 为:I like cat
sb2 为:I like dog

五、改错题

设 s1 和 s2 为 String 类型的字符串,s3 和 s4 为 StringBuffer 类型的字符串,下列哪个语句或表达式不正确？
```
s1 = "Hello World!";
s3 = "Hello World!";
String s5 = s1+s2;
StringBuffer s6 = s3+s4;
String s5 = s1-s2;
s1<=s2
char c = s1.charAt(s2.length());
s4.setCharAt(s4.length(),'y');
```
答案：语句或表达式不正确的有：
```
s3 = "Hello World!";
StringBuffer s6 = s3+s4;
String s5 = s1-s2;
s1<=s2
```

4.3 上机实验

一、实验目的与意义

1. 掌握数组、字符串的定义和使用方法；
2. 熟悉数组的排序、查找等算法；
3. 掌握字符数组的使用方法。

二、实验内容

1. 编写数组的排序程序；
2. 编写折半查找法的程序；
3. 编程实现：产生一个1~12之间的随机整数，并根据该随机整数的值，输出对应月份的英文名称；
4. 编程实现：建立包含10个字符串数据的一维数组，每个字符串数据的格式为"MM/DD/YY"，例如，06/23/12，将每个日期采用类似"23th June 2012"的格式输出。

三、实验要求

1. 在数组的排序程序中，随机产生20个整数，对其按照升序进行排列，并对排序前后的数组按照每行5个数的方式输出。
2. 通过命令行输入一个数，在排序后的数组中，采用折半查找法查找该数在数组中的位置。如果查找到该数，输出信息：XXX：Y。其中XXX代表待查找数，Y代表该数在数组中的位置（下标）。
3. 用赋初值的方法，将1~12月的英文月份名赋给数组元素，根据所产生的随机整数值，输出对应的数组元素值。
4. 用赋初值的方法，将10个日期格式的字符串数据赋予数组元素，然后按照指定格式输出。

4.4 程序代码

1. 排序程序

```java
public class Sort {
    public static void bubble(int a[])
    {
        int count=a.length,i;
        for(i=0;i<count;i++)
            for(int j=count-1;j>i;j--)
                if(a[j]<a[j-1]){
                    int temp=a[j];
                    a[j]=a[j-1];
                    a[j-1]=temp;
                }
    }
    public static void main(String[] args) {
        final int l=20;
        int a[]=new int[l];
        for(int k=0;k<a.length;k++)
            a[k]=(int)(100*Math.random());
        System.out.println("Before sort:");
        for(int i=0;i<a.length;i++){
            System.out.print(a[i]+"   ");
            if((i+1)%5==0)
                System.out.println();
        }
        System.out.println("After sort:");
        bubble(a);
        for(int i=0;i<a.length;i++)
        {
            System.out.print(a[i]+"   ");
            if((i+1)%5==0)
                System.out.println();
        }
    }
}
```

2. 查找程序

```java
public class Bisearch1 {
    public static int bisearch(int a[],int n) {
        int low=0,high=a.length-1,mid=(low+high)/2;
        System.out.println("After sort:");
        for(int k=0;k<a.length;k++)
```

```java
        System.out.print(a[k]+" ");
     System.out.println();
     if(n>a[high]||n<a[low])return -1;
     while(low<=high)
     {
        if(a[low]==n)return low;
        else if(a[high]==n )return high;
        else if(a[mid]==n) return mid;
        else
        {
           if(n>mid)
           {low=mid+1;high=high-1;}
           else
           {high=mid-1;low=low+1;}
        }
     }
     return -1;
}
public static void sort(int a[])
{
   int count=a.length,i;
   for(i=0;i<count;i++)
      for(int j=count-1;j>i;j--)
         if(a[j]<a[j-1]){
            int temp=a[j];
            a[j]=a[j-1];
            a[j-1]=temp;
         }
}
public static void main(String args[]){
   final int l=10;
   int a[]=new int[l];
   int n,i;
   for(i=0;i<args.length-1;i++)
      a[i]=(int)(100*Math.random());
   n=Integer.parseInt(args[0]);
   sort(a);
   int place=bisearch(a,n);
   if(place! =-1)
      System.out.println(n+":"+place);
   else
      System.out.println(n+"is not found.");
}
```

3. 随机程序

```java
public class Month {
    public static void main(String[] args) {
        String months[] = {"January","February","March","April","May","June","July",
            "August","September","October","November","December"};
        int n = (int)(1+Math.random()*12);
        System.out.println(months[n-1]);
    }
}
```

4. 日期程序

```java
public class DateForm {
    public static void main(String[] args) {
        String months[] = {"January","February","March","April","May","June","July",
            "August","September","October","November","December"};
        String suffix[] = {"1st","2nd","3rd"};
        String date[] = {"06/25/04","07/03/07","08/20/08","08/24/08"};
        for(int i=0;i<date.length;i++) {
            String year="20"+date[i].substring(6,8);
            int month=Integer.parseInt(date[i].substring(0,2));
            int day=Integer.parseInt(date[i].substring(3,5));
            if(day<1||day>31)
                System.out.println("Day is illegal");
            if(month>0&&month<13) {
                if(day>0&&day<4)
                    System.out.println(date[i]+":"+suffix[day-1]+" "+months[month-1]+" "+year);
                else
                    System.out.println(date[i]+":"+day+"th "+months[month-1]+" "+year);
            }
            else
                System.out.println("Month is illegal!");
        }
    }
}
```

第5章 类和对象

5.1 典型例题解析

【例5.1】 运行下面的程序,并改正其错误。理解面向对象的程序设计方法的基本概念。

```
class SumXY
{
   public int x,y;
   public SumXY( ){x=a; y=b;}
}
public class Expriment4_1
{
   public static void main(String[ ] args) //主函数
   {
      int total;
      SumXY num=new SumXY(24,65);
      total=num.x+num.y;
      System.out.println("add="+total);
   }
}
```

解析:构造方法出错。构造方法参数列表中无参数,构造实例对象时无法调用相应的构造方法。

答案:应将原题中的构造方法修改为:public SumXY(int a,int b) {x=a;y=b;}

运行结果如图5.1所示。

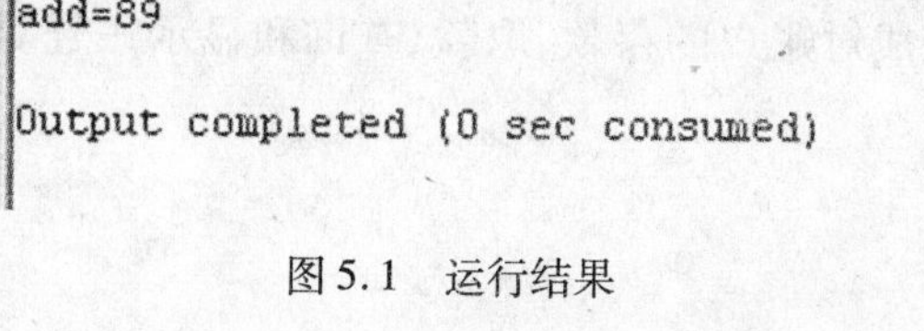

```
---------- 运行 ----------
add=89

Output completed (0 sec consumed)
```

图5.1 运行结果

【例5.2】 请在横线处填上正确的语句,并上机调试至正确。

```java
public class Expriment5_2
{
    int num1;
    double num2;
    public Expriment5_2(int number1,double number2)
    {
        num1=number1;
        num2=number2;
    }
    public static void main(String[] args)
    {
        int a=23;
        double b=4.5;
        Expriment4_2 temp=new Expriment4_2(a,b);
        System.out.println("num1="+             );//输出显示 temp 对象 num1 的值
        System.out.println("num2="+             );//输出显示 temp 对象 num2 的值
    }
}
```

解析：实例对象调用当前的成员变量，调用格式为：对象.成员变量名

答案：

temp.num1

temp.num2

运行结果如图 5.2 所示。

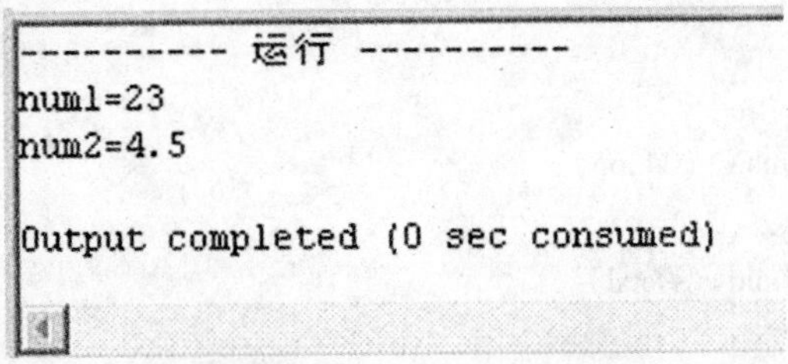

```
---------- 运行 ----------
num1=23
num2=4.5

Output completed (0 sec consumed)
```

图 5.2 运行结果

【例 5.3】 定义一个类实现银行账户的概念，包括的属性有"账号"和"存款余额"，包括的方法有"存款"、"取款"、"查询余额"和"显示账号"。编写一测试类，创建两个不同的账户类的对象，并分别完成存款、取款、查询余额、显示账号等操作。

解析：创建一个银行账户 Count 类，并定义两个成员变量 ID(账号)和 yue(余额)并有对当前成员变量设置和获取的成员方法。并创建 4 个成员方法，分别是 cun( )，qu( )，chaxun( )和 xianshizhanghao( )用来完成对账户的存款、取款、查询和显示账号等操作。最后通过测试类，创建实例对象完成操作。

```java
class Count
{
    private int ID;
    private float yue;
    public Count(int ID,float yue)
```

```java
        {
            this.ID=ID;
            this.yue=yue;
        }
        public void cun(float moneycun)
        {
            yue=yue+moneycun;
        }
        public void qu(float moneyqu)
        {
            yue=yue-moneyqu;
            System.out.println("取款后还有 "+yue+" 元");
        }
        public void chaxun()
        {
            System.out.println("余额为："+yue    );
        }
        public void xianshizhanghao()
        {
            System.out.println("ID 为 "+ID    );
        }
    }

public class test4_3
{
    public static void main(String[] args)
    {
        Count c1=new Count(001,1000);
        Count c2=new Count(002,2000);
        c1.cun(20);
        c1.xianshizhanghao();
        c1.chaxun();
        c2.cun(100);
        c2.qu(35);
        c2.xianshizhanghao();
        c2.chaxun();
    }
}
```

运行结果如图 5.3 所示。

```
---------- 运行 ----------
ID为 1
余额为: 1020.0
取款后还有 2065.0 元
ID为 2
余额为: 2065.0
Output completed (0 sec consumed)
```

图 5.3　运行结果

5.2　课后习题解答

一、选择题

1～5 ACADD　6～10 CDADC　11～13 CBD

二、填空题

1. class　2. 构造方法　3. 非 static　4. 形参　5. 值传递　6. 局部变量　本方法　7. 修饰符　8. 定义　9. 类名　10. 任何

三、程序填空

①return x;

②return y;

③p. setXY(3,4);

④Rectangle r=new Rectangle();

⑤System. out. println("r's length is :"+r. getLength()+" r's width is :"
+r. getWidth()+"　r's position is :"+r. getPosition(). getX()
+""+r. getPosition(). getY());

四、简答题

1. 如何对对象进行初始化?

答:对对象初始化有两种方法:“对象名. 数据成员=值”和“利用构造方法初始化”。

2. 静态数据成员与非静态数据成员有何不同?

答:静态数据成员与非静态数据成员有以下不同:

(1)用 static 修饰符修饰的数据成员是静态数据成员,它是属于类的数据成员,而无 static 修饰符修饰的数据成员是非静态数据成员,它是属于类的对象的数据成员。

(2)静态数据成员被保存在类的公共存储区中,因此,一个类的任何对象访问它时,存取到的都是相同的数值。而非静态数据成员是保存在某个对象的存储区中。

(3)静态数据成员可以通过类名. 静态数据成员或对象名. 静态数据成员方法来访问,而非答:静态数据成员只能通过对象名. 非静态数据成员方式来访问。

3. 静态成员方法与非静态成员方法有何不同？

答：静态成员方法与非静态成员方法有以下不同：

（1）静态成员方法属于整个类，非静态成员方法在对象被创建时具体属于某个对象。

（2）静态成员方法可以使用类名.静态成员方法或对象名.静态成员方法来访问，而非静态成员方法只能通过对象名.非静态成员方法方式来访问。

（3）静态方法只能访问静态数据成员；非静态方法可以访问静态或非静态数据成员。

（4）静态方法只能访问静态方法；非静态方法可以访问静态或非静态方法。

（5）静态方法不能被覆盖，非静态方法可以被覆盖。

4. final 数据成员和成员方法有什么特点？

答：用 final 修饰符修饰的数据成员被限定为最终数据成员。最终数据成员可在声明时进行初始化，也可通过构造方法赋值，但不能在程序的其他部分赋值，它的值在程序的整个执行过程中是不能被改变的。用 final 修饰符修饰的数据成员是标识符常量。用 final 修饰符说明常量时，需要说明常量的数据类型并指出常量的具体值。若有多个对象，而某个数据成员是常量，最好将次常量声明为 static，即用 static final 两个修饰符修饰，这样可节省空间。

用 final 修饰的方法称为最终方法。类的子类不能覆盖父类的方法，即不能再重新定义与父类的方法同名的方法，而仅能使用从父类继承来的方法。使用 final 修饰的方法，是为防止任何继承类修改次方法，保证了程序的安全性和正确性。

5. 类的修饰符有什么作用？

答：类的修饰符用于说明对它的访问限制，常用的有以下几种：

（1）public：公共类，可被其他包中的类访问。

（2）abstract：说明该类为抽象类。

（3）final：最终类，没有子类。

（4）默认：可被同一个包中的类使用。

五、读程序写结果

1. 程序的运行结果（图 5.4）是：

3+5=8

4+7+8=19

3.4+2.2=5.6

1.1+2.2+3.3=6.6

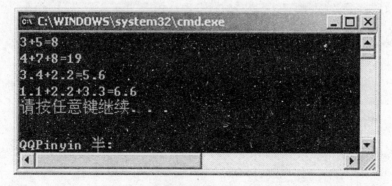

图 5.4 读程序写结果程序运行截图

2. 程序的运行结果(图 5.5)是：
p1's x is 0　p1's y is 0
p2's x is 1　p2's y is 1
p3's x is 0　p3's y is 0

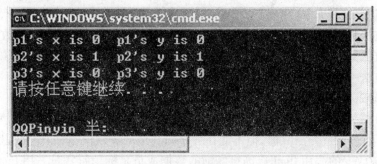

图 5.5　读程序写结果程序运行截图

3. 程序的运行结果(图 5.6)是：
OutClass
Inclass
Inclass

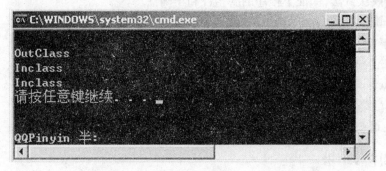

图 5.6　读程序写结果程序运行截图

4. 程序的运行结果(图 5.7)是：
中国
中国
中国

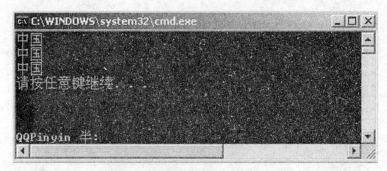

图 5.7　读程序写结果程序运行截图

六、程序设计

1. 程序功能:编写一个学校类,其中包括成员变量 scoreLine(录取分数线)和对该变量进行设置和获取的方法;编写一个学生类,它的成员变量有考生的 name(姓名),id(考号),intgretResult(综合成绩),sports(体育成绩)。还有获取和设置学生的综合成绩和体育成绩的方法;编写一个录取类,它的一个方法用于判断学生是否符合录取条件。其中录取条件为:综合成绩在录取分数线上,或体育成绩在 96 分以上并且综合成绩大于 300 分。该类中的 main 方法建立若干个学生对象,对符合录取条件的学生,输入其信息"被录取",运行程序并查看结果。

```
class School{
    private float scoreLine;
    public School(float scoreLine){ this. scoreLine=scoreLine ;}
    public void setScoreLine(float scoreLine){this. scoreLine=scoreLine ;}
    public float getScoreLine(){ return scoreLine ;}
}
class Students{
    private String name;
    private int id;
    private float intgretResult;
    private float sports;
    public Students(int id,String name,float intgretResult,float sports){
        this. name=name;
        this. id=id;
        this. intgretResult=intgretResult;
        this. sports=sports;
    }
    public void setIntgretReaulit(float intgretResult){
        this. intgretResult=intgretResult ;
    }
    public void setSports(float sports){this. sports=sports;}
    public float getIntgretResult(){ return intgretResult ;}
    public float getSports(){ return sports ;}
    public String toString(){return "姓名:"+name+"\t";}
}
public class LuQu{
    Students stu;
    School sch;
    public LuQu(Students stu, School sch){
        this. stu=stu;
        this. sch=sch;
    }
    public void LuQuFou(){
        if((stu. getIntgretResult()>sch. getScoreLine())||
```

```java
            (stu.getSports()>=96)&&(stu.getIntgretResult()>300))
            System.out.println(stu.toString()+"被录取");
        else
            System.out.println(stu.toString()+"很遗憾");
    }
    public static void main(String args[]){
        Students stu[] = new Students[3];
        stu[0] = new Students(001,"张平",270,76);
        stu[1] = new Students(002,"赵娜",315,60);
        stu[2] = new Students(003,"李磊",301,97);
        School sch1 = new School(315);
        for(int i=0;i<stu.length;i++){
            LuQu lq=new LuQu(stu[i],sch1);
            lq.LuQuFou();
        }
    }
}
```

2. 定义一个名为 MyRectangle 的矩形类，类中有 4 个私有的整型域，分别是矩形的左上角坐标(xUp,yUp)和右下角坐标(xDown,yDown)；类中定义没有参数的构造方法和有 4 个 int 参数的构造方法，用来初始化类对象。类中还有以下方法：getW()——计算矩形的宽度；getH()——计算矩形的高度；area()——计算矩形的面积；toString()——把矩形的宽、高和面积等信息作为字符串返回。编写应用程序使用 MyRectangle 类。

```java
class MyRectangle {
    private int xUp,yUp,xDown,yDown;
    public MyRectangle(){xUp=0;yUp=0;xDown=0;yDown=0;}
    public MyRectangle(int xUp,int yUp,int xDown,int yDown){
        this.xUp=xUp;this.yUp=yUp;
        this.xDown=xDown; this.yDown=yDown;
    }
    public int getW(){return xDown-xUp;}
    public int getH(){return yDown-yUp;}
    public int area(){return getW() * getH();}
    public String toString(){return "矩形的宽:"+getW()+"\n 矩形的高:"
            +getH()+"\n 矩形的面积:"+area();}
}
public class Text5_1 {
    public static void main(String[] args) {
        MyRectangle rect=new MyRectangle(10,20,30,60);
        System.out.println(rect.toString());
    }
}
```

5.3 上机实验

一、实验目的

1. 熟练掌握类、对象的概念以及对事物的抽象;
2. 熟练掌握成员变量和方法的概念以及构造方法的概念;
3. 理解面向对象的程序设计方法。

二、实验要求

1. 编写一个体现面向对象思想的程序;
2. 编写一个创建对象和使用对象的方法程序;
3. 编写不同成员变量修饰方法的程序。

三、实验内容

1. 创建对象:new 构造函数(构造方法参数列表);
2. 使用修饰符;
常用的修饰符如下:
[public] [private] [protected] [package] [static] [final] [transient] [volatile]
不同修饰符的访问控制权限如表 5.1 所示。

表 5.1 修饰符访问权限

修饰符	类	子类	包	所有类和包
public	√	√	√	√
private	√			
protected	√	√	√	
package	√		√	

3. 方法中参数传递的练习。

5.4 程序代码

一、创建对象和修饰符使用

(1) 程序功能:通过两个类 StaticDemo、Exp5_1 说明静态变量/方法与实例变量/方法的区别。

(2) 编写类文件 Exp5_1.java,程序源代码如下:

```
class StaticDemo {
static int x;
int y;
public static int getX() {
```

```java
        return x;
    }
    public static void setX(int a) {
        x = a;
    }
    public int getY() {
        return y;
    }
    public void setY(int b) {
        y = b;
    }
}

public class Exp5_1 {
    public static void main(String[] args) {
        System.out.println("静态变量 x = "+StaticDemo.getX());
        System.out.println("实例变量 y = "+StaticDemo.getY()); //非法,编译时将出错
        StaticDemo a = new StaticDemo();
        StaticDemo b = new StaticDemo();
        a.setX(1);
        a.setY(2);
        b.setX(3);
        b.setY(4);
        System.out.println("静态变量 a.x = "+a.getX());
        System.out.println("实例变量 a.y = "+a.getY());
        System.out.println("静态变量 b.x = "+b.getX());
        System.out.println("实例变量 b.y = "+b.getY());
    }
}
```

(3)对上面的源程序进行编译,会出现如图 5.8 所示的出错提示。

```
<terminated> Exp5_1 [Java Application] D:\eclipse\EasyEclipse Desktop Java 1.3.1.1\jre\bin\javaw.exe (2011-11-30 下午08:02:29)
Exception in thread "main" java.lang.Error: Unresolved compilation problem:
    Cannot make a static reference to the non-static method getY() from the type StaticDemo

    at Exp5_1.main(Exp5_1.java:20)
```

图 5.8　程序运行结果

(4)将源程序中的出错语句删除或使用解释符//隐藏起来,例如,

//System.out.println("实例变量 y = "+StaticDemo.getY());

(5)重新编译并运行该程序,结果如图 5.9 所示。

static 关键字声明的成员变量/方法被视为类的成员变量/方法,而不把它当作实例对象的成员变量/方法。换句话说,静态变量/方法是类固有的,可以直接引用,其他成员变量/方法仅仅被声明生成实例对象后才存在,才可以被引用。基于这样的事实,也把静态变量/方法称为类变量/方法,非静态变量称为实例变量/方法。从实验结果可以得出以下几点结论:

①类的静态变量可以直接引用,而非静态变量则不行。类的静态变量相当于某些程序语言的全局变量。

```
<terminated> Exp5_1 [Java Application]
静态变量x=0
静态变量a.x=3
实例变量a.y=2
静态变量b.x=3
实例变量b.y=4
```

图 5.9　程序运行结果

② 静态方法只能使用静态变量,不能使用实例变量。因为对象实例化之前,实例变量不可用。

③ 类的静态变量只有一个版本,所有实例对象引用的都是同一个版本。

④ 对象实例化后,每个实例变量都被制作了一个副本,它们之间互不影响。

二、方法中参数传递的练习

在其他语言中,函数调用或过程调用时参数有传值调用和传地址调用之分。在 Java 中,方法中的参数传递可以分为传值调用或对象方法调用等方式。传值调用即传递的参数是基本数据类型,调用方法时在方法中将不能改变参数的值,这意味着只能使用它们。对象调用是指先调用对象,再调用对象的方法,这种方式可以修改允许存取的成员变量。所以,如果不想改变参数的值,可以采用传值调用的方法。如果想改变参数的值,可采用对象调用的方法,间接修改参数的值。

1. 编写一个传值调用的程序文件 Exp5_2.java

(1) 程序功能:程序首先给整型变量 x 和 y 赋初值 10,然后使用传值调用方式调用方法 ff1 对 x 和 y 做乘方及输出 x 和 y 的乘方值,最后再输出 x 和 y 的乘方值。

(2) 程序源代码如下:

```java
class Exp5_2{
    public static void main(String[] args) {
        int x=10, y=10;
        ff1(x, y);
        System.out.println("x="+x+", y="+y);
    }
    static void ff1(int passX, int passY) {
        passX = passX * passX;
        passY = passY * passY;
        System.out.println("passX="+passX+", passY="+passY);
    }
}
```

(3) 编译 Exp5_2.java,其运行结果如图 5.10 所示。

```
<terminated> Exp5_2 [Java Application]
passX=100, passY=100
x=10, y=10
```

图 5.10　程序运行结果

(4)分析其运行结果。这个程序没有实现预期的结果,原因是 ff1 方法采用了传值调用。调用 ff1 方法时,将产生两个参数 passX 和 passY,x 和 y 的值被传递给这两个参数。尽管在方法中计算了参数的平方,但从 ff1 方法返回后,参数消失,此时 x 和 y 的值仍是初值。

2. 编写一个调用对象方法的程序文件 Exp5 _ 3. java

(1)程序功能:通过调用对象的方法在方法调用后修改了成员变量的值。

(2)Exp5 _ 3.java 程序源代码如下:

```java
public class Exp5 _ 3{
    public static void main(String[ ] args) {
        Power p=new Power( );
        p. ff2(10,10);
        System. out. println("方法调用后 x ="+p. x+", y ="+p. y);
    }
}
class Power{
    int x=10, y=10;
    void ff2(int passX, int passY) {
    System. out. println("初始时 x ="+x+", y ="+y);
    x=passX * passX;
    y=passY * passY;
    System. out. println("方法调用中 x ="+x+", y ="+y);
    }
}
```

(3)编译 Exp5 _ 3.java,其运行结果如图 5.11 所示。

```
<terminated> Exp5_3 [Java Application]
初始时x=10, y=10
方法调用中 x=100, y=100
方法调用后 x=100, y=100
```

图 5.11 程序运行结果

第6章 类的继承和多态

6.1 典型例题解析

一、类的继承性练习

1. 进一步理解继承的含义

新类可从现有的类中产生,保留现有类的成员变量和方法,并可根据需要对它们加以修改;新类还可添加新的变量和方法。这种现象就称为类的继承。当建立一个新类时,不必写出全部成员变量和成员方法。只要简单地声明这个类是从一个已定义的类继承下来的,就可以引用被继承类的全部成员。被继承的类称为父类或超类(superclass),这个新类称为子类。通常要对子类进行扩展,即添加新的属性和方法。这使得子类要比父类大,但更具特殊性,代表着一组更具体的对象。继承的意义就在于此。

2. 创建公共类 LX3_7_P

(1)编写程序文件 LX3_7_P.java,源代码如下:

```
public class LX3_7_P
{
    protected String xm;  //具有保护修饰符的成员变量
    protected int xh;
    void setdata(String m,int h)  //设置数据的方法
    {
        xm = m;
        xh = h;
    }
    public void print()  //输出数据的方法
    {
        System.out.println(xm+","+xh);
    }
}
```

(2)编译 LX3_7_P.java,产生类文件 LX3_7_P.class。
(3)创建继承的类。
程序功能:通过 LX3_7_P 类产生子类 LX3_8,其不仅具有父类的成员变量 xm(姓名)、

xh(学号),还定义了新成员变量xy(学院)、xi(系)。在程序中调用了父类的print方法,同时可以看出子类也具有该方法。

编写LX3_8.java程序,源代码如下:

```java
class LX3_8 extends LX3_7_P{
    protected String xy;
    protected String xi;
    public static void main(String args[])
    {
        LX3_7_P p1=new LX3_7_P();
        p1.setdata("帅零",12321);
        p1.print();
        LX3_8 s1=new LX3_8();
        s1.setdata("郭丽娜",12345);   //调用父类的成员方法
        s1.xy="经济管理学院";   //访问本类的成员变量
        s1.xi="信息管理系";   //访问本类的成员变量
        s1.print();
        System.out.print(s1.xm+","+s1.xy+","+s1.xi);
    }
}
```

(4)编译并运行程序,其结果如图6.1所示。

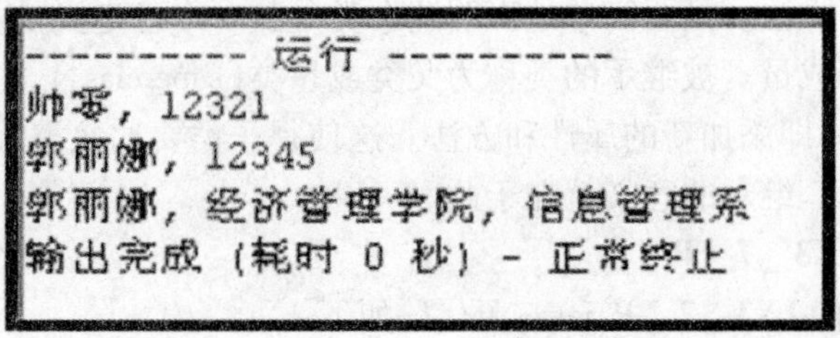

```
---------- 运行 ----------
帅零,12321
郭丽娜,12345
郭丽娜,经济管理学院,信息管理系
输出完成 (耗时 0 秒) - 正常终止
```

图6.1 运行结果

3. 了解成员变量的隐藏方式

所谓隐藏是指子类重新定义了父类中的同名变量,在子类Line中重新定义了x为x1,y为y1,隐藏了父类Point中的两个成员变量x和y。子类执行自己的方法时,操作的是子类的变量,子类执行父类的方法时,操作的是父类的变量。在子类中要特别注意成员变量的命名,防止无意中隐藏了父类的关键成员变量,这有可能给程序带来麻烦。

4. 了解成员方法的覆盖方式

(1)方法覆盖的定义与作用。通过继承子类可以继承父类中所有可以被子类访问的成员方法,但如果子类的方法与父类方法同名,则不能继承,此时称子类的方法覆盖了父类的方法,简称为方法覆盖(override)。方法覆盖为子类提供了修改父类成员方法的能力。例如,子类可以修改层层继承下来的Object根类的toString方法,让它输出一些更有用的信息。下面的程序显示了在子类Circle中添加toString方法,用来返回圆半径和圆面积信息。

(2)编写覆盖Object类toString方法的程序文件LX3_9.java,源代码如下:

```java
class Circle {
    private int radius;
    Circle(int r) {
```

```
      setRadius(r);
    }
    public void setRadius(int r){
      radius=r;
    }
    public int getRadius(){
      return radius;
    }
    public double area(){
      return 3.14159*radius*radius;
    }
    public String toString(){
      return "圆半径:"+getRadius()+" 圆面积:"+area();
    }
}
public class LX3_9{
    public static void main(String args[]){
      Circle c=new Circle(10);
      System.out.println("\n"+c.toString());
    }
}
```

(3)编译并运行程序,其结果如图6.2所示。

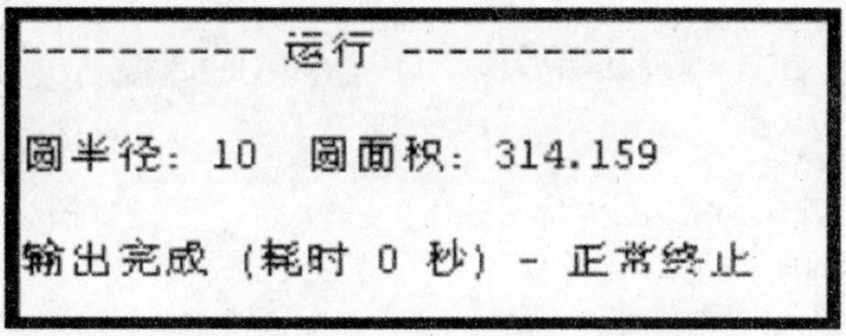

```
---------- 运行 ----------

圆半径: 10  圆面积: 314.159

输出完成 (耗时 0 秒) - 正常终止
```

图6.2 运行结果

(4)程序结构分析。

程序添加了 toString 方法并修改了它的返回值。由于 toString 和继承下来的 Object 类的方法名相同,返回值类型相同,因此覆盖了超类 Object 中的 toString 方法。

方法覆盖时要特别注意:

用来覆盖的子类方法应和被覆盖的父类方法保持同名、相同的返回值类型,以及相同的参数个数和参数类型。

5. this,super 和 super( )的使用

(1)程序功能:说明 this,super 和 super( )的用法。程序首先定义 Point(点)类,然后创建点的子类 Line(线),最后通过 LX3_10 类输出线段的长度。

程序中通过 super(a,b)调用父类 Point 的构造方法为父类的 x 和 y 赋值。在子类 Line 的 setLine 方法中,因为参数名和成员变量名相同,为给成员变量赋值,使用 this 引用,告诉编译器是为当前类的成员变量赋值。在 length 和 toString 方法中使用父类成员变量时,使用 super 引用,告诉编译器使用的是父类的成员变量。

(2) 使用 this，super 和 super( ) 的程序文件 LX3_10.java，源代码如下：

```java
class Point {
    protectedint x, y;
    Point(int a, int b) {
        setPoint(a, b);
    }
    public voidsetPoint(int a, int b) {
        x=a;
        y=b;
    }
}
class Line extends Point {
    protectedint x, y;
    Line(int a, int b) {
        super(a, b);
        setLine(a, b);
    }
    public voidsetLine(int x, int y) {
        this.x=x+x;
        this.y=y+y;
    }
    public double length() {
        int x1=super.x, y1=super.y, x2=this.x, y2=this.y;
        returnMath.sqrt((x2-x1)*(x2-x1)+(y2-y1)*(y2-y1));
    }
    public StringtoString() {
        return "直线端点:["+super.x+","+super.y+"]["+
        x+","+y+"]直线长度:"+this.length();
    }
}
public class LX3_10{
    public static void main(Stringargs[]) {
        Line line=new Line(50, 50);
        System.out.println("\n"+line.toString());
    }
}
```

(3) 编译并运行程序，结果如图 6.3 所示。

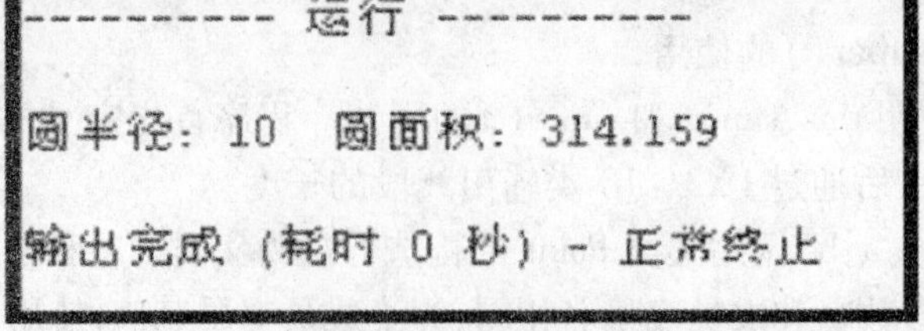

```
---------- 运行 ----------

圆半径: 10   圆面积: 314.159

输出完成 (耗时 0 秒) - 正常终止
```

图 6.3 运行结果

二、类的多态性练习

1. 理解类的多态性

类的继承发生在多个类之间,而类的多态只发生在同一个类上。在一个类中,可以定义多个同名的方法,只要确定它们的参数个数和类型不同。这种现象称为类的多态。多态使程序简洁,为程序员带来很大便利。在 OOP 中,当程序要实现多个相近的功能时,就给相应的方法起一个共同的名字,用不同的参数代表不同的功能。这样,在使用方法时不论传递什么参数,只要能被程序识别就可以得到确定的结果。类的多态性体现在方法的重载(overload)上,包括成员方法和构造方法的重载。

2. 方法的重载

方法重载是让类以统一的方式处理不同类型数据的一种手段。Java 的方法重载,就是类中可以创建多个方法,它们具有相同的名字,但具有不同的参数和不同的定义。调用方法时通过传递给它们的不同个数和类型的参数未决定具体用哪个方法。

3. 构造方法的重载

构造方法的名称和类同名,没有返回类型。构造方法不能直接调用,只能由 new 操作符调用,主要用来完成对象的初始化。重载构造方法的目的是提供多种初始化对象的能力,使程序员可以根据实际需要选用合适的构造方法来初始化对象。

(1)程序功能:编写构造方法 RunDemo 的重载程序文件 LX3_12,源代码如下:

```
classRunDemo {
  private StringuserName, password;
  RunDemo( ) {
    System. out. println("全部为空!");
  }
  RunDemo( String name) {
    userName = name;
  }
  RunDemo( String name, String pwd) {
    this( name);
    password = pwd;
    check( );
  }
  void check( ) {
    String s = null;
    if ( userName! = null)
      s = "用户名:"+userName;
    else
      s = "用户名不能为空!";
    if ( password! = "12345678")
      s = s+"口令无效!";
    else
      s = s+"口令:* * * * * * * *";
    System. out. println( s);
```

        }
    }
    public class LX3_12 {
        public static void main(String[ ] args) {
            newRunDemo( ) ;
            newRunDemo("刘新宇") ;
            newRunDemo(null ,"邵丽萍") ;
            newRunDemo("张驰","12345678") ;
        }
    }

（2）编译并运行程序,结果如图 6.4 所示。

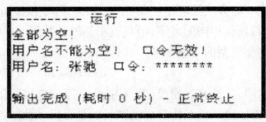

图 6.4 运行结果

三、编程题

定义点类 Point，扩展到线 Line 类和圆 Circle 类，这是三个公共类,不能放在同一个文件中。它们都没有输出语句,如果运行看不到什么结果。

```
public class Point {
protected int x, y;
    Point(int a, int b) {setPoint(a, b);}
    public void setPoint(int a, int b) {
        x = a;
        y = b;
    }
    public int getX( ) {return x;}
    public int getY( ) {return y;}
}

public class Line extends Point {
    protected int x, y, endX, endY;
    Line(int x1, int y1, int x2, int y2) {setLine(x1, y1, x2, y2);}
    public void setLine(int x1, int y1, int x2, int y2) {
        x = x1;
        y = y1;
        endX = x2;
        endY = y2;
    }
    public int getX( ) {return x ;}
```

```
publicint getY() {return y;}
publicint getEndX() {return endX;}
publicint getEndY() {return endY;}
public double length() {
returnMath.sqrt((endX-x)*(endX-x)+(endY-y)*(endY-y));
   }
}
public class Circle extends Point {
  protectedint radius;
  Circle(int a, int b, int r) {
    super(a, b);
    setRadius(r);
  }
  public void setRadius(int r) {radius=r;}
  public int getRadius() {return radius;}
  public double area() {return 3.14159*radius*radius;}
}
```

Point 的成员

x, y	// 受保护的成员变量，代表点的坐标
Point	// 点的构造方法
setPoint	// 设定点的坐标值的方法
getX, getY	// 返回坐标 x 和 y 的值的方法

Line 的成员

x, y, endX, endY	// 子类受保护的成员变量，代表线的两个端点坐标
Line	// 线的构造方法
setLine	// 设定线的两个端点坐标值的方法
getX, getY	// 返回起点坐标 x 和 y 的值的方法
getEndX, getEndY	// 返回终点坐标 endX 和 endY 的值的方法
length	// 返回线的长度的方法
x, y	// 继承父类的受保护成员变量，但被子类隐藏
setPoint	// 继承父类的方法
getX, getY	// 继承父类的方法，但被子类覆盖

Circle 的成员

radius	// 子类受保护的成员变量，代表圆的半径
Circle	// 圆的构造方法
setRadius	// 设定半径值的方法
getRadius	// 返回半径值的方法
area	// 返回圆面积的方法
x, y	// 继承父类的受保护成员变量
setPoint	// 继承父类的方法
getX, getY	// 继承父类的方法

6.2 课后习题解答

一、选择题

1~5 CDBCA

二、填空题

1. 派生 基 2. 复用 3. 小于 4. 多 5. private 6. extends 7. 当前对象 8. 直接父类 9. 内部类

三、简答题

1. 什么是多态？面向对象程序设计为什么要引入多态的特性？使用多态有什么优点？

答：多态是指程序中同名的不同方法共存的情况。多态是面向对象程序设计的又一个特性。我们知道，面向过程的程序设计中，过程或函数各具有一定的功能，它们之间是不允许重名的；而面向对象程序设计中，则要利用这种多态来提高程序的抽象性，突出 Java 语言的继承性。面向对象的程序中多态的情况有多种，可以通过子类对父类方法的覆盖实现多态，也可以利用重载在同一个类中定义多个同名的不同方法。

多态的特点大大提高了程序的抽象程度和简捷性，更重要的是它最大限度地降低了类和程序模块之间的耦合性，提高了类模块的封闭性，使得它们不需了解对方的具体细节，就可以很好地共同工作。这个优点，对程序的设计、开发和维护都有很大的好处。

2. Overload 和 Override 的区别是什么？

答：方法的重写 Overriding 和重载 Overloading 是 Java 多态性的不同表现。重写 Overriding 是父类与子类之间多态性的一种表现，重载 Overloading 是一个类中多态性的一种表现。

如果在子类中定义某方法与其父类有相同的名称和参数，我们说该方法被重写(Overriding)。子类的对象使用这个方法时，将调用子类中的定义，对它而言，父类中的定义如同被"屏蔽"了。如果在一个类中定义了多个同名的方法，它们或有不同的参数个数或有不同的参数类型，则称为方法的重载(Overloading)。Overloaded 的方法是可以改变返回值的类型。

四、程序设计

1. import java.awt.*;
```
import java.applet.*;
public class ChongZai extends Applet
{
    int add(int a,int b)
    {
        return a+b;
    }
    int add(int a,int b,int c)
    {
        return a+b+c;
```

```
    }
    public void paint(Graphics g)
    {
        g.drawString("sum is :"+add(2,5),5,10);
        g.drawString("sum is :"+add(7,8,9),5,30);
    }
}
```

程序运行结果如图 6.5 所示。

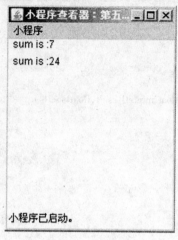

图 6.5　运行结果

2. class a
```
{
    int x=5;
}
public class 数据成员的隐藏
{
    int x=778;
    public static void main(String[ ]args)
    {
        int m,n;
        a s=new a();
        数据成员的隐藏 ss=new 数据成员的隐藏();
        m=s.x;
        n=ss.x;
        System.out.println("m :"+m);
        System.out.println("n :"+n);
    }
}
```

程序运行结果如图 6.6 所示。

3. class Student
```
{
String name;
```

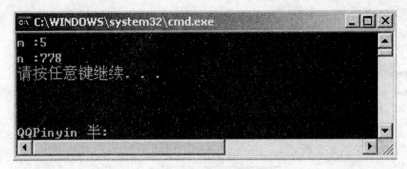

图 6.6 运行结果

```
int id;
floatintgretResult;
float sports;
public Student(Stringname, int id, float intgretResult, float sports)
{
   this.name = name;
   this.id = id;
   this.intgretResult = intgretResult;
   this.sports = sports;
}
public floatgetIntgretResult( )
{
   return intgretResult;
}
public floatgetSports( )
{
   return sports;
}
}
class School
{
floatscoreLine;
public School(floatscoreLine)
{
   this.scoreLine = scoreLine;
}
public floatgetScoreLine( )
{
   return scoreLine;
}
public voidsetScoreLine(float scoreLine)
{
   this.scoreLine = scoreLine;
}
```

```java
}
public class LuQu
{
    Student stu;
    School sch;
    public LuQu(Student stu,School sch)
    {
        this.stu=stu;
        this.sch=sch;
    }
    public boolean luQuFou()
    {
        if(((stu.intgretResult>sch.scoreLine)||(stu.sports>96&&stu.intgretResult>300)))
        {
            return true;
        }
        else//否则
        {
            return false;
        }
    }
    public static void main(String[] args) //主函数
    {
        Student stu1=new Student("zhangsan",001,270,76);
        School sch=new School(250);
        LuQu lq=new LuQu(stu1,sch);
        if(lq.luQuFou())
        {
            System.out.println(stu1.name);
            System.out.println("恭喜你!");
        }
        else
        {
            System.out.println(stu1.name);
            System.out.println("很抱歉!");
        }
    }
}
```

程序运行结果如图6.7所示。

图6.7 运行结果

6.3 上机实验

一、实验目的与意义

1. 掌握 OOP 方式进行程序设计的方法；
2. 了解类的继承性和多态性的作用。

二、实验内容

1. 阅读如下所示的 3 个 Java 类的定义，分析它们之间的关系，写出运行结果。

2. 假定根据学生的 3 门学位课程的分数决定其是否可以拿到学位，对于本科生，如果 3 门课程的平均分数超过 60 分即表示通过，而对于研究生，则需要平均超过 80 分才能够通过。根据上述要求，请完成以下 Java 类的设计：

(1) 设计一个基类 Student 描述学生的共同特征。

(2) 设计一个描述本科生的类 Undergraduate，该类继承并扩展 Student 类。

(3) 设计一个描述研究生的类 Graduate，该类继承并扩展 Student 类。

(4) 设计一个测试类 StudentDemo，分别创建本科生和研究生这两个类的对象，并输出相关信息。

3. 假定要为某个公司编写雇员工资支付程序，这个公司有各种类型的雇员(employee)，不同类型的雇员按不同的方式支付工资：

(1) 经理(manager)——每月获得一份固定的工资。

(2) 销售人员(salesman)——在基本工资的基础上每月还有销售提成。

(3) 一般工人(worker)——按他每月工作的天数计算工资。

根据上述要求试用类的继承和相关机制描述这些功能，并编写一个 JavaApplication 程序，演示这些类的用法。(提示：应设计一个雇员类(employee)描述所有雇员的共同特性，这个类应该提供一个计算工资的抽象方法 ComputeSalary( )，使得可以通过这个类计算所有雇员的工资。经理、销售人员和一般工人对应的类都应该继承这个类，并重新定义计算工资的方法，进而给出它的具体实现。)

三、实验要求

1. 编写体现类的继承性(成员变量，成员方法，成员变量隐藏)的程序；

2. 编写体现类多态性(成员方法重载,构造方法重载)的程序;
3. JDK1.6 与 eclipse 开发环境。

6.4 程序代码

1. 阅读如下所示的 3 个 Java 类的定义,分析它们之间的关系,写出运行结果。

```
classSuperClass {
    int x;
    SuperClass() {
        x=3;
        System.out.println("in SuperClass : x="+x);
    }
    void doSomething() {
        System.out.println("in SuperClass.doSomething()");
    }
}
classSubClass extends SuperClass {
    int x;
    SubClass() {
    super();//调用父类的构造方法
        x=5;//super() 要放在方法中的第一句
        System.out.println("in SubClass :x="+x);
    }
    void doSomething() {
        super.doSomething();//调用父类的方法
        System.out.println("in SubClass.doSomething()");
        System.out.println("super.x="+super.x+" sub.x="+x);
    }
}
public class Inheritance {
    public static void main(String args[]) {
        SubClass subC=new SubClass();
        subC.doSomething();
    }
}
```

运行结果:
    inSuperClass: x=3
    inSubClass: x=5
    inSuperClass.doSomething()
    inSubClass.doSomething()
    super.x=3 sub.x=5

2. 假定根据学生的 3 门学位课程的分数决定其是否可以拿到学位,对于本科生,如果 3 门课程的平均分数超过 60 分即表示通过,而对于研究生,则需要平均超过 80 分才能够通过。根

据上述要求,请完成以下 Java 类的设计:
(1)设计一个基类 Student 描述学生的共同特征。
(2)设计一个描述本科生的类 Undergraduate,该类继承并扩展 Student 类。
(3)设计一个描述研究生的类 Graduate,该类继承并扩展 Student 类。
(4)设计一个测试类 StudentDemo,分别创建本科生和研究生这两个类的对象,并输出相关信息。

```java
class Student{
    private String name;
    private int classA,classB,classC;
    public Student(String name,int classA,int classB,int classC){
        this.name=name;
        this.classA=classA;
        this.classB=classB;
        this.classC=classC;
    }
    public String getName(){
        return name;
    }
    public int getAverage(){
        return (classA+classB+classC)/3;
    }
}

class UnderGraduate extends Student{
    public UnderGraduate(String name,int classA,int classB,int classC){
        super(name,classA,classB,classC);
    }
    public void isPass(){
        if(getAverage()>=60)
            System.out.println("本科生"+getName()+"的三科平均分为:"+getAverage()+",可以拿到学士学位。");
        else
            System.out.println("本科生"+getName()+"的三科平均分为:"+getAverage()+",不能拿到学士学位。");
    }
}

class Graduate extends Student{
    public Graduate(String name,int classA,int classB,int classC){
        super(name,classA,classB,classC);
    }
    public void isPass(){
        if(getAverage()>=80)
            System.out.println("研究生"+getName()+"的三科平均分为:"+getAverage()+",可以拿到硕士学位。");
```

```
    else
        System.out.println("研究生"+getName()+"的三科平均分为:"+getAverage()+",
        不能拿到硕士学位。");
    }
}
public classStudentDemo{
    public static void main(String[] args){
        UnderGraduate s1=new UnderGraduate("Tom",55,75,81);
        Graduate s2=new Graduate("Mary",72,81,68);
        s1.isPass();
        s2.isPass();
    }
}
```

运行结果:
本科生 Tom 的三科平均分为:70,可以拿到学士学位。
研究生 Mary 的三科平均分为:73,不能拿到硕士学位。

3. 假定要为某个公司编写雇员工资支付程序,这个公司有各种类型的雇员(employee),不同类型的雇员按不同的方式支付工资:

(1)经理(manager)——每月获得一份固定的工资。
(2)销售人员(salesman)——在基本工资的基础上每月还有销售提成。
(3)一般工人(worker)——按他每月工作的天数计算工资。

根据上述要求试用类的继承和相关机制描述这些功能,并编写一个 JavaApplication 程序,演示这些类的用法。(提示:应设计一个雇员类(employee)描述所有雇员的共同特性,这个类应该提供一个计算工资的抽象方法 ComputeSalary( ),使得可以通过这个类计算所有雇员的工资。经理、销售人员和一般工人对应的类都应该继承这个类,并重新定义计算工资的方法,进而给出它的具体实现。)

```
abstract class Employee{
    private String name;
    public Employee(String name){
        this.name=name;
    }
    public StringgetName(){
        return name;
    }
    public abstract doublecomputeSalary();
}
class Manager extends Employee{
    private doublemonthSalary;
    public Manager(Stringname,double monthSalary){
        super(name);
        this.monthSalary=monthSalary;
    }
    public doublecomputeSalary(){
```

```java
        return monthSalary;
    }
}
class Salesman extends Employee{
    private double baseSalary;
    private double commision;
    private int qualtities;
    public Salesman(String name,double baseSalary,double commision,int qualtities){
        super(name);
        this.baseSalary=baseSalary;
        this.commision=commision;
        this.qualtities=qualtities;
    }
    public double computeSalary(){
        return baseSalary+commision * qualtities;
    }
}
class Worker extends Employee{
    private double dailySalary;
    private int days;
    public Worker(String name,double dailySalary,int days){
        super(name);
        this.dailySalary=dailySalary;
        this.days=days;
    }
    public double computeSalary(){
        return dailySalary * days;
    }
}
public class EmployeeDemo{
    public static void main(String args[]){
        Manager e1=new Manager("张三",10000);
        Salesman e2=new Salesman("李四",2000,50.4,63);
        Worker e3=new Worker("王五",79.5,28);
        System.out.println("经理"+e1.getName()+"的月工资为:"+e1.computeSalary());
        System.out.println("销售人员"+e2.getName()+"的月工资为:"+e2.computeSalary());
        System.out.println("工人"+e3.getName()+"的月工资为:"+e3.computeSalary());
    }
}
```

运行结果：
经理张三的月工资为:10000.0
销售人员李四的月工资为:5175.2
工人王五的月工资为:2226.0

第7章 接口、抽象类与包

7.1 典型例题解析

一、了解并使用 Java 的系统包

包是类和接口的集合。利用包可以把常用的类或功能相似的类放在一个包中。Java 语言提供系统包,其中包含了大量的类,可以在编写 Java 程序时直接引用它们。所有 JavaAPI 包都以"java."开头,以区别用户创建的包。接口解决了 Java 不支持多重继承的问题,可以通过实现多个接口达到与多重继承相同的功能。处理程序运行时的错误和设计程序同样重要,只有能够完善处理运行时出错的程序,软件系统才能长期稳定地运行,异常处理的作用就是如何处理程序运行时出错的问题。

二、创建并使用自定义包

1. 自定义包的声明方式

\<package\> \<自定义包名\>

声明包语句必须添加在源程序的第一行,表示该程序文件声明的全部类都属于这个包。

2. 创建自定义包 Mypackage

在存放源程序的文件夹中建立一个子文件夹 Mypackage。例如,在"E:\javademo"文件夹之中创建一个与包同名的子文件夹 Mypackage(E:\javademo\Mypackage),并将编译过的 class 文件放入该文件夹中。注意:包名与文件夹名大小写要一致。再添加环境变量 classpath 的路径,例如,D:\java\jdk1.6\lib; E:\javademo。

3. 在包中创建类

(1) YMD.java 程序功能:在源程序中,首先声明使用的包名 Mypackage,然后创建 YMD 类,该类具有计算今年的年份,可以输出一个带有年月日的字符串的功能。

(2) 编写 YMD.java 文件,源代码如下:

```
package Mypackage; //声明存放类的包
import java.util.*; //引用 java.util 包
public class LX4_1_YMD {
    private int year,month,day;
    public static void main(String[] arg3){}
```

```java
    public LX4_1_YMD(int y,int m,int d){
      year=y;
      month=(((m>=1) & (m<=12)) ? m : 1);
      day=(((d>=1) & (d<=31)) ? d : 1);
    }
    public LX4_1_YMD(){
      this(0,0,0);
    }
    public static int thisyear(){
      return Calendar.getInstance().get(Calendar.YEAR);//返回当年的年份
    }
    public int year(){
      return year;//返回年份
    }
    public String toString(){
      return year+"-"+month+"-"+day;//返回转化为字符串的年-月-日
    }
}
```

(3) 编译 LX4_1_YMD.java 文件,然后将 LX4_1_YMD.class 文件存放到 Mypackage 文件夹中。

4. 编写使用包 Mypackage 中 LX4_1_YMD 类的程序

(1) LX4_2.java 程序功能:给定某人姓名与出生日期,计算该人年龄,并输出该人姓名、年龄、出生日期。程序使用了 LX4_1_YMD 的方法来计算年龄。

(2) 编写 LX4_2.java 程序文件,源代码如下:

```java
import Mypackage.LX4_1_YMD;//引用 Mypackage 包中的 LX4_1_YMD 类
public class LX4_2
{
    private String name;
    private LX4_1_YMD birth;
    public static void main(String args[])
    {
      LX4_2 a=new LX4_2("张驰",1990,1,11);
      a.output();
    }
    public LX4_2(String n1,LX4_1_YMD d1)
    {
      name=n1;
      birth=d1;
    }
    public LX4_2(String n1,int y,int m,int d)
    {
      this(n1,new LX4_1_YMD(y,m,d));//初始化变量与对象
    }
```

```
    public int age()  //计算年龄
    {
        return LX4_1_YMD.thisyear() - birth.year();  //返回当前年与出生年的差,即年龄
    }
    public void output()
    {
        System.out.println("姓名:"+name);
        System.out.println("出生日期:"+birth.toString());
        System.out.println("今年年龄:"+age());
    }
}
```

(3) 编译并运行程序,结果如图 7.1 所示。

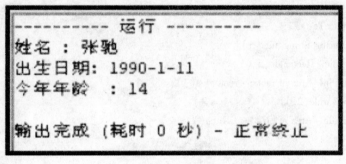

图 7.1　运行结果

三、使用接口技术

1. 接口的定义与作用

接口可以看作是没有实现的方法和常量的集合。接口与抽象类相似,接口中的方法只是作了声明,而没有定义任何具体的操作方法。使用接口是为了解决 Java 语言中不支持多重继承的问题。

(1) 定义一个接口 Shape2D,利用它来实现二维的几何形状类 Circle 和 Rectangle 面积计算编写实现接口的程序文件。

(2) 源代码如下:

```
interface Shape2D{  //定义 Shape2D 接口
    final double pi=3.14;  //数据成员一定要初始化
    public abstract double area();  //抽象方法,不需要定义处理方式
}

class Circle implements Shape2D{
    double radius;
    public Circle(double r){  //构造方法
        radius=r;
    }
    public double area(){
        return (pi * radius * radius);
```

```
    }
  }

class Rectangle implements Shape2D{
  int width,height;
  public Rectangle(int w,int h){//构造方法
    width=w;
    height=h;
  }
  public double area(){
    return (width * height);
  }
}
public class InterfaceTester {
public static void main(String args[]){
  Rectangle rect=new Rectangle(5,6);
  System.out.println("Area of rect="+rect.area());
  Circle cir=new Circle(2.0);
  System.out.println("Area of cir="+cir.area());
  }
}
```

7.2 课后习题解答

一、选择题

1~5 DCDDC 6~9 CABC

二、填空题

1.抽象 2.abstract 3.常量数据成员 4.构造抽象

三、简答题

1.抽象类和接口的区别是什么？

答：抽象类在 Java 语言中表示的是一种继承关系，一个类只能使用一次继承关系。但是，一个类却可以实现多个接口。

在抽象类中可以有自己的数据成员，也可以有非抽象的成员方法，而在接口中，只能够有静态的不能被修改的数据成员（也就是必须是 static final 的，不过在接口中一般不定义数据成员），所有的成员方法都是抽象的。

抽象类和接口所反映出的设计理念不同。其实抽象类表示的是"is-a"关系，接口表示的是"like-a"关系。

实现抽象类和接口的类必须实现其中的所有方法。抽象类中可以有非抽象方法，接口中则不能有实现方法。

接口中定义的变量默认是 public static final 型,且必须给其赋初值,所以实现类中不能重新定义,也不能改变其值。

抽象类中的变量默认是 friendly 型,其值可以在子类中重新定义,也可以重新赋值。接口中的方法默认都是 public 或 abstract 类型的。

2. 为什么不能将类同时声明为 abstract 和 final?

答:一个类不能同时声明为 abstract 和 final,这是因为一个类同时使用 abstract 和 final 描述没有意义。当一个类声明为 abstract 时,不能从这个类创建对象。这个类型表示该类描述的对象过于一般,需要将抽象类子类化,然后用它的非抽象子类创建对象。但是一个最终类 final 不能被子类化,结果意味抽象的最终类是无用的。

3. 什么是接口?为什么要定义接口?接口与类有什么异同?

答:接口是用来实现类间多重继承功能的结构。它定义了若干个抽象方法和常量用以实现多重继承的功能。

Java 语言不支持多重继承,只支持单重继承(只有一个直接父类)。然而在解决实际问题的程序设计中仅靠单重继承尚不能解决更复杂的问题。为了使 Java 程序的类层次结构更加合理,更符合实际问题的需要,我们把用于完成特定功能的若干属性组织成相对独立的属性集合。这种属性的集合就是接口。

定义接口与定义类非常相似。实际上完全可以把接口理解成为一个特殊的类,接口是由常量和抽象方法组成的特殊类。一个类只能有一个父类,但是它可以同时实现若干个接口。这种情况下如果把接口理解成特殊的类,那么这个类利用接口实际上就获得了多个父类,即实现了多重继承。与类定义相仿,声明接口时也需要给出访问控制符,不同的是接口的访问控制符只有 public 一个。用 public 修饰的接口是公共接口,可以被所有的类和接口使用,而没有 public 修饰符的接口则只能被同一个包中的其他类和接口利用。接口也具有继承性。定义一个接口是可以通过 extends 关键字声明该新接口是某个已经存在的父类接口的派生接口,它将继承父接口的所有属性和方法。与类的集成不同的是一个接口可以有一个以上的父接口,它们之间用逗号分隔,形成父接口列表。新接口将继承所有父接口中的属性和方法。

四、程序设计

1. 编写一个 JavaApplication 程序,该程序有一个 Pointer 类,它包含横坐标 x 和纵坐标 y 两个属性,再给 Pointer 定义两个构造方法和一个打印点坐标的方法 Show。定义一个圆 Circle 类,它继承 Pointer 类(它是一个点,圆心(Center)),除此之外,还有属性半径 Radius,再给圆定义两个构造方法:一个打印圆的面积的方法 PrintArea 和一个打印圆中心、半径的方法 Show。

分析:Pointer 定义两个构造方法:一个可以是无参的构造函数,另一个可以是有参数的。构造函数的功能可以是只对它的两个属性 x 和 y 初始化。Circle 类继承 Pointer 类,要注意构造函数如何实现继承。另外,因为两个类中都有 show 方法,也要考虑如何调用的问题。现将 Pointer 类和 Circle 类的定义提供如下:

```
class Pointer
{   int x;
    int y;
public Pointer( )
{   x=0;
```

```java
        y=0;
    }
    public Pointer(int a,int b)
    {SetXY(a,b);}
    public int GetX()
    {return x;}
    public int GetY()
    {return y;}
    public void SetXY(int a,int b)
    {x=a;y=b;}
    public void Show()
    {
        System.out.println("点:"+"("+x+","+y+")" );
    }
}

class Circle extends Pointer
{final double PI=3.1415926;
int Radius;
Circle()
{Radius=5;}
Circle(Pointer Center,int Radius)
{   super(Center.GetX(),Center.GetY());
    this.Radius=Radius;
}
public void PrintArea()
{   double area=PI*Radius*Radius;
    System.out.println("*********************");
    System.out.println("面积:"+area );
    System.out.println("*********************");
}
    public void Show()
    {   super.Show();
        System.out.println("半径:"+ Radius );
    }

    public static void main(String args[])
    {
        Pointer p=new Pointer();
        Circle c=new Circle(p,3);
        c.PrintArea();
        c.Show();
    }
}
```

运行结果如图7.2所示。

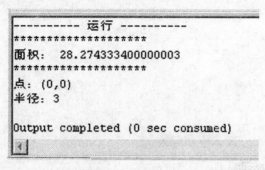

图 7.2　运行结果

2. 编写一测试类,对其进行编译、运行。结果如何? 如去掉语句"super.show();",再看看运行结果。理解程序中重载和多态性的运用。

完成以下步骤要求:

(1) 设计一个表示二维平面上点的类 Point,包含有表示坐标位置的 protected 类型的成员变量 x 和 y,获得和设置 x 和 y 值的 public 方法。

(2) 设计一个表示二维平面上圆的类 Circle,它继承自类 Point,还包含表示圆半径的 protected 类型的成员变量 r、获取和设置 r 值的 public 方法、计算圆面积的 public 方法。

(3) 设计一个表示圆柱体的类 Cylinder,它继承自类 Circle,还包含有表示圆柱体高的 protected 类型的成员变量 h、获得和设置 h 值的 public 方法、计算圆柱体体积的 public 方法。

(4) 建立若干个 Cylinder 对象,输出其轴心坐标、半径、高及其体积的值。

```
class Point
{
    protected float x,y;
    public Point( )
    {
    }
    public Point(int x,int y)
    {
        this.x=x;
        this.y=y;
    }
    public void setXY(int x,int y)
    {
        this.x=x;
        this.y=y;
    }
    public float getX( )
    {
        return x;
    }
    public float getY( )
    {
        return y;
```

```java
    }
}
class Circle extends Point
{
    protected float r;
    public Circle()
    {
    }
    public Circle(float x,float y,float r)
    {
        this.x=x;
        this.y=y;
            this.r=r;
    }
    public float getR()
    {
        return r;
    }
    public void setR(float r)
    {
        this.r=r;
    }
    public float area()
    {
        return 3.14f*r*r;
    }
}

class Cylinder extends Circle
{
    protected float h;
    public Cylinder(float x,float y,float r,float h)
    {
        this.x=x;
        this.y=y;
        this.r=r;
        this.h=h;
    }
    public void setH(float h)
    {
        this.h=h;
    }
    public float getH()
    {
```

```
      return h;
    }
    public float area()
    {
      return 3.14f * r * r;
    }
    public float tiji()
    {
      return area() * h;
    }
  }

public class test5_2
{
  public static void main(String[] args)  //主函数
  {
    Cylinder cy1 = new Cylinder(1,2,3,10);
    System.out.println("轴心坐标为:"+cy1.getX()+"  "+cy1.getY());
    System.out.println("半径为:"+cy1.getR());
    System.out.println("圆柱体高为:"+cy1.getH());
    System.out.println("圆柱体积为:"+cy1.tiji());
  }
}
```

运行结果如图7.3所示。

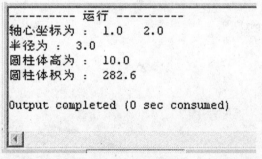

图7.3　运行结果

3.学校中有老师和学生两类人,而在职研究生既是老师又是学生,对学生的管理和对教师的管理在他们身上都有体现。

(1)设计两个信息管理接口 StudentInterface 和 TeacherInterface。其中,StudentInterface 接口包括 setFee 方法和 getFee 方法,分别用于设置和获取学生的学费;TeacherInterface 接口包括 setPay 方法和 getPay 方法,分别用于设置和获取教师的工资。

(2)定义一个研究生类 Graduate,实现 StudentInterface 接口和 TeacherInterface 接口,它定义的成员变量有 name(姓名),sex(性别),age(年龄),fee(每学期学费),pay(月工资)。

(3)创建一个姓名为"张三"的研究生,统计他的年收入和学费,如果收入减去学费不足2 000元,则输出"provide a loan"(需要贷款)信息,否则输出"够自己花了"。

```java
interface StudentInterface
{
void setFee(float fee);
float getFee();
}
interface TeacherInterface
{
void setPay(float pay);
float getPay();
}

class Graduate implements StudentInterface,TeacherInterface
{
String name;
char sex;
int age;
float fee,pay;
public Graduate(String name,char sex,int age,float fee,float pay)
{
    this.name=name;
    this.sex=sex;
    this.age=age;
    this.fee=fee;
    this.pay=pay;
}
public void setPay(float pay)
{
    this.pay=pay;
}
public float getPay()
{
    return pay;
}
public void setFee(float fee)
{
    this.fee=fee;
}
public float getFee()
{
    return fee;
}
}
public class test5_3
{
```

```
public static void main(String[] args)
{
    Graduate 张三 = new Graduate("张三",'男',23,4500,7800);
    if(张三.getPay()-张三.getFee()<2000)
    {
        System.out.println("provide a loan");
    }
    else
    {
        System.out.println("够自己花了!");
    }
}
```

运行结果如图7.4所示。

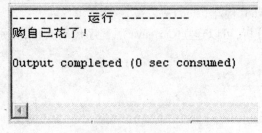

图7.4 运行结果

7.3 上机实验

一、实验目的与意义

1. 正确理解 Java 中包的基本概念；
2. 基本掌握包的创建以及包中类的引用；
3. 正确理解接口的构成及作用。

二、试验内容

1. 创建一个简单的汽车(car)组件包(其中包括车身车架合成、发动机合成、控制组件合成)；
2. 根据接口的特点将汽车的控制组件类 ControlModule 改为接口实现。

三、实验要求

1. JDK1.5 与 eclipse 开发工具；
2. 根据包、类之间的关系，设计类(父类、子类或一般类)将它们放入包中，完成类程序的编写与调试，掌握包中类的引用方法；
3. 根据接口的特点，编写接口程序；完成常用异常处理程序的编写和调试，掌握异常处理在程序中的应用方式。

7.4 程序代码

1. 创建一个简单的汽车(car)组件包(其中包括车身车架合成、发动机合成、控制组件合成)。

(1) 汽车组件简要分析。大家知道汽车是一个抽象的概念,因为汽车的种类繁多,功能各异。但不管何类汽车都包括如下组件:车身车架(含驾驶室)、发动机、控制组件(变速器、驱动桥、转向制动系统等)。因此一辆汽车可以看成是若干部件的组合。如果我们分别把它们划分为汽车部件类,可以简单地描述如下:

车身车架(CarBody)类:属性可以有型号、颜色、轮胎个数等;动作有开门(open)、关门(close)等。

发动机(Engine)类:属性可以有类型、功率等;动作有启动(start)、关闭(close)等。

控制组件(ControlCompModule)类:动作有前进(forward)、倒退(backward)、加速(quicken)、减速(slowDown)、停车(stop)等。

(2) 根据如上简单分析,可以建立 Car.java 源程序代码文件,分别定义各组件类,源程序参考代码如下:

```java
package car;
/*定义车身车架类*/
class CarBody
{
    String type;     //声明类型
    String color;    //声明颜色
    int tire;        //声明轮胎数
    public void open()
    {
        System.out.println("车门已打开!");
    }
    public void close()
    {
        System.out.println("车门已关闭!");
    }
    public CarBody(String type,String color,int tire)
    {
        this.type=type;
        this.color=color;
        this.tire=tire;
    }
}
/*定义发动机类*/
class Engine
{
    String model;    //声明发动机型号
```

```java
    int power;        //声明功率
    public void start()
    {
        System.out.println("发动机已发动!");
    }
    public void close()
    {
        System.out.println("发动机已关闭!");
    }
    public Engine(String model,int power)
    {
        this.model=model;
        this.power=power;
    }
}
/* 以下定义控制系统类 */
class ControlModule
{
    public void forward(int speed)
    {
        System.out.println("汽车以"+speed+"速度开始向前行进...");
    }
    public void backward(int speed)
    {
        System.out.println("请注意:正在以"+speed+"速度倒车!");
    }
    public void stop()
    {
        System.out.println("已经停车!");
    }
}
```

(3) 编译 Car.java 源程序代码,在 car 包中生成各个类的字节码文件,供其他应用程序使用。我们可以查看一下,在当前的文件夹下,编译系统自动创建了 car 文件夹,并把编译生成的类文件 CarBody.class,Engine.class,ControlModule.class 放在了 car 文件夹下。

在完成上边的操作之后,作为本实验的作业完成下边的自由练习:

编写测试程序 TestCar.java,组合一个汽车对象,测试汽车部件类的功能。

完成该示例的方法步骤如下:

(1) 这是一个测试类的程序,在测试程序中,我们需要组合一个汽车对象,然后分别执行对象的各个部件的行为方法。测试程序的参考代码如下:

```java
package car;
public class TestCar
{
    CarBody carBody;    //声明车体构架
```

```
    Engine  engine;      //声明发动机
    ControlModule CM;    //声明控制组件
    public TestCar(String type,String color,int tire,String model,int power)
    {
        carBody=new CarBody(type,color,tire);  //创建车体构架成员对象
        engine=new Engine(model,power);        //创建发动机成员对象
        CM=new ControlModule();                //创建控制组件成员对象
    }
    public static void main(String [] args)
    {
        TestCar car=new TestCar("奔马 2000","褐色",4,"N-Ⅳ-B5",220);//创建汽车对象
        car.carBody.open();      //打开车门
        car.carBody.close();     //关闭车门
        car.engine.start();      //启动发动机
        car.CM.forward(30);      //向前行进
        car.CM.stop();           //停止前进
        car.CM.backward(1);      //倒车
        car.engine.close();      //关闭发动机
        car.carBody.open();      //打开车门
        car.carBody.close();     //关闭车门
        System.out.println("测试已完成");
    }
}
```

（2）在建立 TestCar.java 源程序文件之后,编译并执行程序,执行结果如图 7.5 所示。

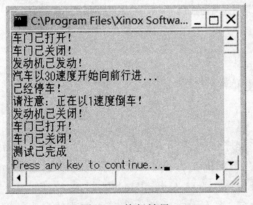

图 7.5　执行结果

（3）程序及结果分析与思考。在测试程序中,我们声明了汽车组件类变量,把它们作为汽车测试类的成员来对待,它们既是本类的成员,也是相对独立的对象,因此在本类的构造方法中创建了这些对象,当本类对象引用时,采用了对象.组件对象.方法的引用方式。

这是一个带包语句的测试类程序,思考如下问题:

① 如果把语句 package car;从程序中去掉,将会发生什么情况? 为什么?

② 将语句 package car;换成:import car.*;可以吗? 为什么?

③ 如果使用语句 import car.*;应该如何重新说明组件类?

2. 根据接口的特点将汽车的控制组件类 ControlModule 改为接口实现。

（1）简要分析。其实，汽车的控制系统是一个较为复杂的系统，这里我们只是对变速器进行了简单的模拟，不论是手工换挡还是自动换挡，其作用都是一样的。假如我们模拟一个换挡器，并为每一挡规定速度，0 挡为停，小于 0 挡为倒车，大于 0 挡为前进，那么只需要定义一个方法，而用参数来确定前进、停止和后退动作就可以了。

（2）根据如上简单分析，可以建立如下的 ControlInterface.java 源程序代码文件：

```
package car;    //放入 car 包中
public interface ControlInterface
{
    public void shiftGear(int speed);    //声明接口方法
}
```

（3）编译程序，生成 ControlInterface.class 文件，并将它放入 car 包中。

（4）程序分析及思考。可以看到接口文件在编译后生成了类文件，因此说接口也是一个特殊的类，接口体中只能包含常量和抽象方法定义，和抽象类相比较，思考如下问题：

① 如果使用抽象类来定义换挡操作，可以吗？

② 分别用抽象类和接口定义相同的内容，能获得相同的效果吗？为什么？

编写测试程序 TestInterface.java，组合一个汽车对象，使用 ControlInterface 接口实现变速控制，测试接口的功能。

完成该示例的方法步骤如下：

（1）这是一个测试类的程序，参照上面测试程序，实现 ControlInterface 接口替代 ControlModule 类，测试程序的参考代码如下：

```
package car;
public class TestInterface implements ControlInterface
{
    CarBody carBody;    //声明车体构架
    Engine   engine;    //声明发动机
    public TestInterface(String type, String color, int tire, String model, int power)
    {
        carBody = new CarBody(type, color, tire);    //创建车体构架成员对象
        engine = new Engine(model, power);    //创建发动机成员对象
    }
    public void shiftGear(int speed)    //实现接口方法
    {
        if(speed>0) System.out.println("汽车以"+speed+"挡速度行进...");
        else if(speed==0) System.out.println("汽车已停止前进!!!");
        else System.out.println("正以"+speed+"挡速度倒车...");
    }
    public static void main(String [] args)
    {
        TestInterface car = new TestInterface("奔马2000","褐色",4,"N-Ⅳ-B5",220);
        //创建汽车对象
        car.carBody.open();    //打开车门
```

```
        car.carBody.close();      //关闭车门
        car.engine.start();       //启动发动机
        car.shiftGear(3);         //向前行进
        car.shiftGear(0);         //停止前进
        car.shiftGear(1);         //倒车
        car.engine.close();       //关闭发动机
        car.carBody.open();       //打开车门
        car.carBody.close();      //关闭车门
        System.out.println("测试已完成");
    }
}
```

（2）在建立 TestInterface.java 源程序文件之后,编译并执行程序。

（3）程序及结果分析与思考。在测试程序中,我们使用接口 ControlInterface 替换了 ControlModule 类的功能,请注意使用它们之间的区别。在既可用类也可用接口来处理解决的问题,使用接口可能要更灵活方便一些。

商场会以各种方式进行促销活动,在购买一定数量的商品之后,可以现场抽奖。编写一个抽奖程序,可以从键盘上输入 0~9 中任意一个字符,如果和系统随机产生的字符相吻合,则显示"恭喜中奖了",否则显示"祝君下次好运!"。

实现该示例的方法步骤如下:

（1）首先简要分析实现该示例需要涉及的问题。要从键盘上输入数据,涉及标准设备的输入问题,需要处理系统可能出现的 I/O 异常问题;要使系统随机产生字符,涉及使用 java.util 类包中的 Random 类对象产生随机数的方法来完成。

（2）根据上述分析,编写处理程序 Bonus.java,参考程序代码如下:

```
import java.util.*;
public class Bonus
{
    public static void main(String[] args) throws java.io.IOException
    {
        Random rd=new Random();
        System.out.print("开始抽奖,请输入0~9之间任意一数:");
        while(true)
        {
            int c1=System.in.read();  //从键盘上输入一个字符
            if(c1<48||c1>57) continue; //如果不是0~9,重新输入,并滤掉回车换行字符
            int n=rd.nextInt(10);     //产生10以内的随机数
            if(n==c1-48) System.out.println("恭喜中奖了!请到1号服务台领取奖品!");
            else System.out.println("祝君下次好运!");
            System.out.print("开始抽奖,请输入0~9之间任意一数:");
        }
    }
}
```

（3）编译 Bonus.java 程序之后,执行程序,执行结果如图 7.6 所示。

（4）程序及程序结果分析。在程序中的 main() 方法说明中使用了 throws java.io.

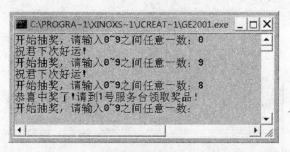

图7.6 执行结果

IOException 子句来抛出 I/O 可能引起的异常,因为这类异常即便进行捕捉,也没有方法对其进行处理。

在使用 System. in. read( )语句接受从键盘上输入的字符时,是以字符的 ASCII 码计值的,我们从键盘上输入的字符"0"~"9"对应的 ASCII 码值为 48~57。

此外,在键盘上输入一个数字然后按"Enter"键,这就包含 3 个字符:所按的数字、回车和换行(对应的 ASCII 码 13 和 10),因此在程序中加入了判别语句:

if(c1<48||c1>57) continue;

它过滤非数字字符(包括回车和换行符)。

第 8 章

异常处理

8.1 典型例题解析

一、异常处理语句格式

异常处理语句格式为：
try ｛…｝   //被监视的代码段，一旦发生异常，则交由其后的 catch 代码段处理
catch（异常类型 e）｛…｝   // 要处理的第一种异常
catch（异常类型 e）｛…｝   // 要处理的第二种异常
…
finally ｛…｝   //最终处理

【例 8.1】 使用 try...catch 语句处理异常的过程

```
public class TC1 {
  public static void main(String[ ] arg3)
{
    System.out.println("这是一个异常处理的例子\n");
    try {
       int i = 10;
       i /= 0;
    }
    catch (ArithmeticException e)
    {
       System.out.println("异常是:"+e.getMessage());
    }
    finally
    {
       System.out.println("finally 语句被执行");
    }
  }
}
```

程序运行结果如图 8.1 所示。

第 8 章 异常处理

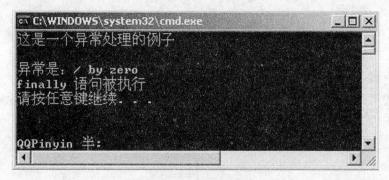

图 8.1 运行结果

【例 8.2】 catch 语句中声明的异常类型不匹配的情况

```
public class TC2 {
   public static void main(String[ ] args) {
      System.out.println("这是一个异常处理的例子");
      try
      {
         int i=10;
         i/=0;
      }
      catch(IndexOutOfBoundsException e)
      {
         System.out.println("异常是:"+e.getMessage());
      }
      finally
      {
         System.out.println("finally 语句被执行");
      }
   }
}
```

程序运行结果如图 8.2 所示。

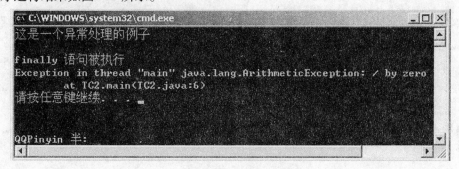

图 8.2 运行结果

【例 8.3】 多个 catch 子句的异常处理

```
public class TC3
{
   public static void main(String[ ] args)
```

```
    {
      try
    {
          int a=args.length;
          System.out.println("\na="+a);
          int b=42/a;
          int c[]={1};
          c[42]=99;
      }
  catch(ArithmeticException e)
    {
        System.out.println("发生了被0除:"+e);
    }
    catch(ArrayIndexOutOfBoundsException e)
    {
        System.out.println("数组下标越界:"+e);
    }
   }
 }
```

程序运行结果如图8.3所示。

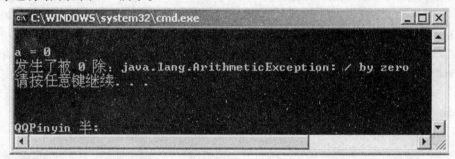

图8.3　运行结果

二、throw 语句格式

throw 语句格式为：

<throw> <new> <异常对象名( )>;

程序会在 throw 语句处立即终止，转向 try...catch 寻找异常处理方法，不再执行 throw 后面的语句。例8.4中使用了 throw 语句主动抛出一个异常。

【例8.4】　throw 语句的使用

```
public class TC4
{
   static void throwProcess()
   {
     try
   {
       throw newNullPointerException("空指针异常");
```

    }
    catch(NullPointerException e)
    {
        System.out.println("\n在 throwProcess 方法中捕获一个"+e.getMessage());
        throw e;
    }
}
public static void main(Stringargs[ ]) {
    try
    {
        throwProcess( );
    }
    catch(NullPointerException e)
    {
        System.out.println("再次捕获:"+e);
    }
}
}
```

程序运行结果如图 8.4 所示。

图 8.4 运行结果

　　throws 用来表明一个方法中可能抛出的各种异常，并说明该方法会抛出异常但不捕获的异常。如果想明确抛出一个 RuntimeException 或自定义异常类，就必须在方法的声明语句中用 throws 子句来表明它的类型，以便通知调用其他方法准备捕获它，这种情况一般需要两个方法来分别处理抛出异常和处理异常。

**1. 抛出异常的方法**

在抛出异常的方法中要使用 throws 子句，throws 子句的格式为：

　　<返回值类型> <方法名><([参数])> < throws> <异常类型> {}

**2. 调用方法处理异常**

　　下面的例子声明了 mathod 方法抛出异常 IllegalAccessException，并且不在 mathod 中捕获它，而在调用 mathod 的 main 方法里捕获它。

【例 8.5】 throws 语句的使用

```
class TC5
{
    static void mathod( ) throws IllegalAccessException
```
```

```
        }
        System.out.println("\n在 mathod 中抛出一个异常");
            throw new IllegalAccessException();
        }
        public static void main(String args[ ]) {
            try
            {
                mathod();
            }
            catch (IllegalAccessException e)
            {
                System.out.println("在 main 中捕获异常:"+e);
            }
        }
    }
```

运行结果如图 8.5 所示。

图 8.5 运行结果

三、由方法抛出异常交由系统处理

对于程序中需要处理的异常,一般编写 try...catch...finally 语句捕获并处理;而对于程序中无法处理必须由系统处理的异常,可以使用 throws 语句在方法中抛出异常交由系统处理。

1. finally 语句

当一个异常被抛出时,程序的执行就不再是连续的了,会跳过某些语句,甚至会由于没有与之匹配的 catch 子句而过早地返回。有时要确保一段代码不管发生什么异常都能被执行是必要的,finally 子句就是用来标识这样一段代码的。即使没有 catch 子句,finally 语句块也会在执行了 try 语句块后立即被执行。每个 try 语句至少都要有一个与之相配的 catch 或 finally 子句。

从一个方法返回到调用它的另外一个方法,或者是通过 return 语句,或者是通过一个没有被捕获的异常,但 finally 子句总是在返回前执行。

【例 8.6】 finally 子句的使用

```
class TC6
{
    static void mathodA()
    {
```

```
        try
        {
            System.out.println("\nmathodA 抛出一个异常");
            throw newRuntimeException( );
        }
        finally
        {
            System.out.println("执行 mathodA 的 finally");
        }
    }
    static voidmathodB( )
    {
        try
        {
            System.out.println("mathodB 正常返回");
            return;
        }
        finally
        {
            System.out.println("执行 mathodB 的 finally");
        }
    }
    public static void main(Stringargs[ ]){
        try
        {
            mathodA( );
        }catch(Exception e){mathodB( );}
    }
}
```

程序运行结果如图 8.6 所示。

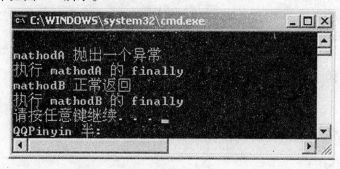

图 8.6　运行结果

2. 编译时对异常情况的检查

（1）可检测的异常。在编译时，编译器分析哪些方法会产生可检测的异常，然后检查方法中的可检测异常的处理部分。如果方法中没有异常处理部分，就要在方法的 throws 子句说明该方法会抛出但不捕获的异常，以告知调用它的其他方法，即将异常上交给调用者处理。

例如,类 Thread 的方法 sleep 定义如下:
public static void sleep (longmillis) throws InterruptedException {…}

(2)不可检测的异常(运行时异常类)。不可检测的异常类是 RuntimeException 及其子类、Error 及其子类,其他的异常类则是可检测的类。标准 JavaAPI 定义了许多异常类,既包括可检测的,也包括不可检测的。由程序员定义的异常类也可以包含可检测类和不可检测类。

3. 创建自己的异常类

自定义异常类型是从 Exception 类中派生的,所以要使用下面的声明语句来创建:
<class> <自定义异常名> <extends> <Exception> {…}

【例 8.7】 创建自定义异常

```
classMyException extends Exception {
    privateint x;
    MyException(int a) {x=a;}
    public StringtoString() {return "MyException";}
}
public class TC7 {
    static voidmathod(int a) throws MyException {
        //声明方法会抛出 MyException
        System.out.println("\t 此处引用 mathod ("+a+")");
        if (a>10) throw newMyException(a);   // 主动抛出 MyException
        System.out.println("正常返回");
    }
    public static void main(Stringargs[ ]) {
        try {
            System.out.println("\n 进入监控区,执行可能发生异常的程序段");
            mathod(8);
            mathod(20);
            mathod(6);
        }
        catch (MyException e) {
            System.out.println("\t 程序发生异常并在此处进行处理");
            System.out.println("\t 发生的异常为:"+e.toString());
        }
        System.out.println("这里可执行其他代码");
    }
}
```

程序的运行结果如图 8.7 所示。

四、典型题分析

1. 简答题

Java 的异常处理机制和方法是什么?

答:(1)异常的抛出:抛出异常首先要生成异常对象,异常可以由 Java 虚拟机和类库中某些类的实例生成,此外还可以在程序中生成异常。声明抛出异常就是在方法声明后通过

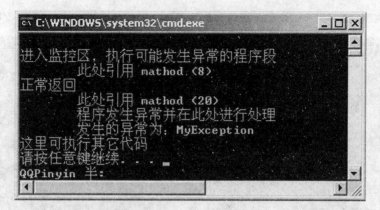

图 8.7　运行结果

throws 子句指明。

（2）异常的捕获：捕获异常就是在某个方法中对某种类型的异常对象提供了相应的处理方法，通常通过 try...catch...finally 语句来实现。其中 try{...}代码块选定捕获异常的范围，try 代码块所限定的语句在执行过程中可能会发生异常对象并抛出该对象。catch 语句用于处理 try 代码中所生成的异常对象，而且 catch 语句可以是多个，每一个 catch 语句带有一个形式参数，参数类型指明能够捕获的异常类型。finally 语句为异常处理提供了一个统一的出口，不管 try 代码中是否发生了异常事件，finally 块中的语句都会被执行。

2. 编程题

（1）编写一个传播和捕获异常的例子。

```
import java.io.*;
class ExceptionDemo1
{
    public static void main( String args[ ] )
    {
        FileInputStream fis=new FileInputStream( "text" );
        int b;
        while( (b=fis.read())!=-1 )
        {
            System.out.print( b );
        }
        fis.close( );
    }
}
```

由于名为"text"的文件在本机硬盘上不存在，所以在编译时，程序出现了 I/O 异常事件，如图 8.8 所示。

（2）关于异常的程序举例。

```
class ExceptionDemo2
{
    public static void main(String [ ]args)
    {
        int a=0;
```

图 8.8 编程题运行结果图

```
        System. out. println(5/a);
    }
}
```

编译：C:\>javac ExceptionDemo2.java
运行：C:\>javaExceptionDemo2
运行结果如图 8.9 所示。
java. lang. ArithmeticException:/by zero at
ExceptionDemo2. main(ExceptionDemo2. java:4)

图 8.9 编程题运行结果图

因为除数不能为 0,所以在程序运行的时候出现了除 0 溢出的异常事件。

(3)编写一个传播和捕获异常的例子,在 main 方法中的 try 块调用 reverse( )函数,如果输出的字符串为空就会抛出一个异常,catch 块会捕获到这个异常,输出字符串"the string is blank";如果输入的字符串不为空,函数就返回倒置的原字符串并输出;finally 块总是要执行的,会输出"well done"。

```
public class Example6 _ 3
{
    public static void main( String[ ]args)
    {
        try{
            System. out. println( reverse("hello")   );
        }
        catch( Exception e)
        {
```

```
                System.out.println("the string was blank");
            }
            finally{
                System.out.println( "well done" );
            }
        }
        public static String reverse(Stringstr) throws Exception{
            if(str.length()==0){
                throw new Exception();
            }
            String reverseStr="";
            for(int i=str.length()-1;i>=0;i--){
                reverseStr+=str.charAt(i);
            }
            return reverseStr;
        }
    }
}
```

程序的运行结果如图 8.10 所示。

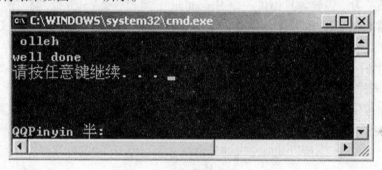

图 8.10　编程题运行结果图

8.2　课后习题解答

一、选择题

1~5 CCDAA

二、简答题

1. 请列举八种常见的可能产生异常的情况。

答：数组下标越界、字符串在取字符时下标越界、除数为 0、类型转换不合适、算数运算结果溢出、运行一个不含方法 main 的 Java 应用程序、非法访问对象的成员域或方法、认为抛出异常(如访问不存在的文件、网络连接不成功等)。

2. 请分别简述关键字 throw 和 throws 的用途,并分析他们之间的差别。

答:(1)throws 用于抛出方法层次的异常,并且直接由一些方法调用异常处理类来处理该异常,所以它常用在方法的后面。例如,

public static void main(String [ ]args) throws SQLException

（2）throw 用于抛出代码层次的异常,常用于方法块里面的代码,常和 try...catch...语句块搭配使用,比 throws 的层次要低。

3.若 try 块未发生异常,try 块执行后,控制转向何处?

答:控制转向 finally 块,若无 finally 块,则继续执行最后一个 catch 块后面的程序。

4.如果发生了一个异常,但没有找到适当的异常处理程序,则会发生什么情况?

答:Java 系统会采用 Java 异常处理机制的预设处理方法来处理。

5.若 try 块中有多个 catch 子句,这些 catch 子句排列次序的不同是否能保证程序执行效果相同?

答:能保证程序执行效果相同。因为每个 catch 块捕获的异常类型不一样。

6.使用 finally 程序块的关键理由是什么?

答:防止强制终止程序,使资源被占用而不释放。finally 程序程序块作为这个组合语句的统一出口,一般用来进行一些"善后"操作,例如,释放资源、关闭文件等。

7.若同时有几个异常处理程序都匹配同一类型的引发对象,则会发生什么情况?

答:编译时会出现有多个 catch 块捕获相同异常的提示,并停止编译。

8.在一个 catch 处理程序中不使用异常类的继承,怎样处理具有相同类型的错误?

答:Java 系统为一组异常只提供一个 Exception 子类和一个 catch 处理程序。当每个异常发生时,该异常对象可以根据不同的实例数据来创建。catch 处理程序将检查该数据,以判断异常的类型。

9.若一个程序引发一个异常,并执行了相应的异常处理程序。但是,在该异常处理程序中又引发了一个同样的异常。这会导致无限循环吗? 为什么?

答:若一个程序引发一个异常,并执行了相应的异常处理程序。但是,在该异常处理程序中又引发了一个同样的异常。这不会导致无限循环。因为当异常处理程序开始运行时,出现异常的 try 块已经终止了,所以此时出现的异常就必须由原 try 块的外层程序来处理。

三、程序设计

1.编写一个程序,包含一个 try 块和两个 catch 块,两个 catch 子句都有能力捕捉 try 块发生的异常。说明两个 catch 子句排列次序不同时程序产生的输出。

```
public class test8_1
{
public static void main( String[ ]args)
{
  int a,b,c;
  a=67;
  b=0;
  try
  {
    int x[ ]=new int[-5];
    c=a/b;
    System.out.println( a+"/"+b+"="+c );
  }
```

```
        catch（ArithmeticException e）
        {
            System.out.println（"b=0："+e.getMessage（））;
        }
        catch（NegativeArraySizeException e）
        {
            System.out.println（"exception："+e.getMessage（））;
            e.printStackTrace（）;
        }
        finally
        {
            System.out.println（"end"）;
        }
    }
}
```

程序运行结果如图 8.11 所示。

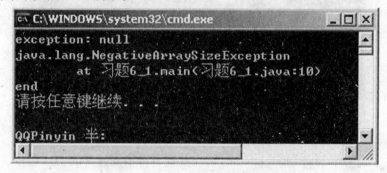

图 8.11　程序设计第一题运行结果图

2. 编写程序，说明引发一个异常是否一定会导致程序终止。
```
public classtest8_2
{
public static void main（String[ ]args）
{
    int a,b,c;
    a=67;
    b=3;
    for（int i=1;i<5 ;i++）
    {
        try
        {
            c=a/（b-i）;
            System.out.println（a+"/"+（b-i）+"="+c）;
        }
        catch（ArithmeticException e）
        {
            if（（b-i）= =0）
```

```
                b=-1;
                System.out.println("b=0: "+e.getMessage());
            }

            finally
            {
                System.out.println("end");
            }
        }
    }
}
```

程序运行结果如图 8.12 所示。

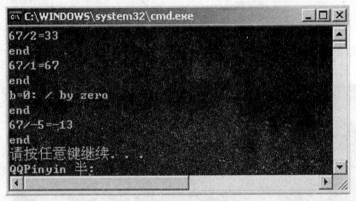

图 8.12 程序设计第二题运行结果图

3. 编写一个说明用 catch(Exception e)可以捕捉各种不同异常的 Java 程序。

```
public class test8_3
{
    public static void main(String[] args)
    {
        int a,b,c;
        a=67;
        b=0;
        try
        {
            int x[]=new int[-5];
            c=a/b;
            System.out.println(a+"/"+b+"="+c);
        }
        catch(Exception e)
        {
            System.out.println("exception "+e.getMessage());
            e.printStackTrace();
        }
        finally
```

```
        }
        System.out.println("end");
    }
}
}
```

程序运行结果如图 8.13 所示。

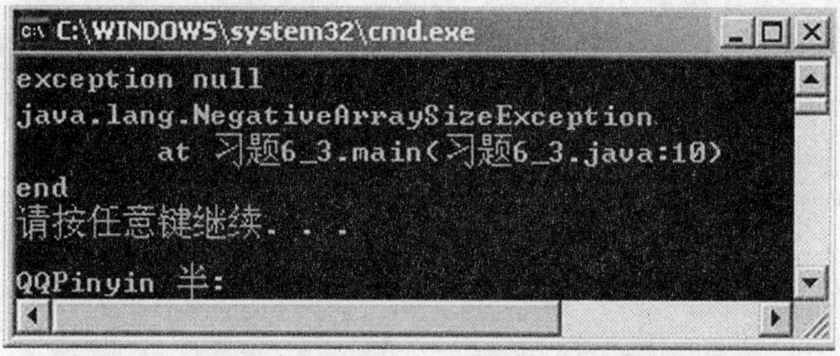

图 8.13　程序设计第三题运行结果图

8.3　上　机　实　验

一、实验目的与意义

1. 理解什么是异常；
2. 掌握 Java 的异常处理机制和方法。

二、实验内容

1. 数组越界；
2. 除 0 异常。

三、实验要求

1. JDK1.5 与 eclipse 开发工具；
2. 了解异常处理的语法；
3. 区分异常和错误。

8.4　程　序　代　码

1. 仔细阅读下面的 Java 语言源程序，自己给出程序的运行结果。

```
importjava.io.*;
public class Experment6_1
{
public static void main(String[]args)    throws IOException
    {
        int [] ko=new int[15];
```

```
    int n,a;
    String x;
    BufferedReader keyin=new BufferedReader(new InputStreamReader(System.in));
    System.out.println("Enter an integer:");
    x=keyin.readLine();
    n=Integer.parseInt(x);
    try
    {
       a=110/n;
       ko[15]=100;
       System.out.println("此描述无法执行");
    }
    catch(ArithmeticException e)
    {
       System.out.println("除数为0的错误!");
    }
    catch(ArrayIndexOutOfBoundsException f)
    {
       System.out.println("数组索引值大于数组长度的错误!");
    }
    System.out.println("执行完catch的描述!!!");
  }
}
```

分别输入12和0,程序的运行结果是什么?为什么?如果把程序中的throws去掉,程序还得如何改?

输入12时,运行结果如图8.14所示。

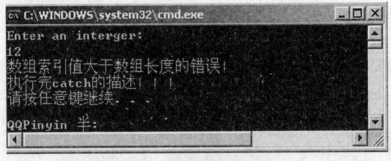

图8.14 运行结果

输入0时运行结果如图8.15所示。

2.仔细阅读下面的Java语言源程序,自己给出程序的运行结果。

```
importjava.io.*;
public class Experment6_2
{
  public Experment6_2()
  {
    try
    {
```

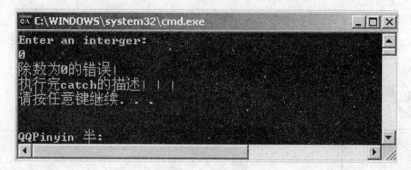

图8.15 运行结果

```
        int a[ ]=new int [2];
        a[4]=3;
        System. out. println( "After handling exception return here?" );
    }
    catch (IndexOutOfBoundsException e)
    {
        System. out. println( "exception msg:"+e. getMessage( ) );
        System. out. println( "exception string:"+e. toString( ) );
        e. printStackTrace( );
    }
    finally
    {
        System. out. println( "--------------------------------" );
        System. out. println( "finally" );
    }
        System. out. println( "No exception?" );//打印
    }
    public static void main( String[ ]args) //主函数
    {
        new Experment6_2( );
    }
}
```

该程序的运行结果是什么？如果想让该程序运行时能将"After handling exception return here?"字符串输出,该如何修改？

程序的运行结果如图8.16所示。

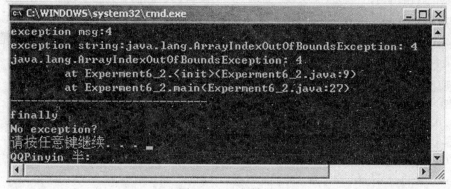

图8.16 运行结果

第 9 章

集合类

9.1 典型例题解析

集合 Collection 接口：Collection 可以存储任何对象组，其元素各自独立，通常拥有相同的套用规则。Set,List 由它派生。基本操作：增加元素 add(Object obj),addAll(Collection c);删除元素 remove(Object obj),removeAll(Collection c);求交集 retainAll(Collection c);访问/遍历集合元素的好办法是使用 Iterator 接口(迭代器用于取代 Enumeration)。

```
Public interface Iterator{
    Public Boolean hasNext();
    Public Object next();
    Public void remove();
}
```

一、set 无重复、无序

Set(集)：集合中的对象不按特定方式排列，并且没有重复对象，它的有些实现类能对集合中的对象按特定方式排列。

set 接口主要有两个实现类 HashSet 和 TreeSet，HashSet 类按照哈希算法来存取集合中的对象，存取速度比较快，HashSet 类还有一个子类 LinkedHashSet 类，不仅实现了哈希算法，而且实现了链表数据结构，TreeSet 类实现了 SortedSet 接口，具有排序功能。HashSet：基于散列表的集，加进散列表的元素要实现 hashCode( )方法以判断是否为同一个对象，无顺序、无重复。

那么，当一个新的对象加入到 Set 集合中，Set 的 add( )方法是如何判断这个对象是否已经存在于集合中的呢？

```
boolean isExists = false;
Iterator it = set.iterator();
while(it.hasNext())
{
    Object oldObject = it.next();
    if(newObject.equals(oldObject))
    {
        sExists = true;
```

```
        break;
    }
}
```

可见,Set 采用对象的 equals()方法比较两个对象是否相等,而不是采用"=="比较运算符,以下程序代码尽管两次调用了 Set 的 add()方法,实际上只加入了一个对象:

```
Set set = new HashSet();
String s1 = new String("hello");
String s2 = new String("hello");
set.add(s1);
set.add(s2);
```

虽然变量 s1 和 s2 实际上引用的是两个内存地址不同的字符串对象,但是由于 s2.equals(s1)的比较结果为 true,因此 Set 认为他们是相等的对象,当第二次调用 Set 的 add()方法时,add()方法不会把 s2 引用的字符串对象加入到集合中。

1. HashSet 类

按照哈希算法来存取集合中的对象,具有很好的存取性能,当 HashSet 向集合中加入一个对象时,会调用对象的 hashCode()方法获得哈希码,然后根据这个哈希码进一步计算出对象在集合中的存放位置。

在 Object 类中定义了 hashCode()和 equals()方法,Object 类的 euqals()方法按照内存地址比较对象是否相等,因此如果 object1.equals(object2)为 true,表明 object1 变量和 object2 变量引用同一个对象。那么 object1 和 object2 的哈希码也应该相同。

如果用户定义的类覆盖了 Object 类的 equals()方法,但是没有覆盖 Object 类的 hashCode()方法,就会导致当 object1.equals(object2)为 true 时,而 object1 和 object2 的哈希码不一定一样,这样使 HashSet 无法正常工作。LinkedHashSet-在 HashSet 中加入了链表数据结构,有顺序。TreeSet-可以排序,需要实现 Comparable 接口,并实现其 compareTo()方法,以排序。

2. TreeSet 类

实现了 SortedSet 接口,能够对集合中的对象进行排序。

例如,

```
Set set = new TreeSet();
set.add(7);
set.add(new Integer(6));
set.add(new Integer(8));
Iterator it = set.iterator();
while(it.hasNext())
{
    System.out.println(it.next());
}
```

输出结果为:6 7 8

当 TreeSet 向集合中加入一个对象时,会把它插入到有序的对象序列中,那么 TreeSet 是如何对对象进行排序的呢? TreeSet 支持两种排序方式:自然排序和客户化排序,默认情况下是自然排序。

在 JDK 中,有一部分类实现了 Comparable 接口,如 Integer、Double 和 String 等,Comparable

接口有一个 compareTo(Object o) 方法,它返回整数类型,对于表达式 x.compareTo(y),如果返回值为 0,表示 x 和 y 相等,如果返回值大于 0,表示 x 大于 y,如果小于 0,表示 x<y。

TreeSet 调用对象的 compareTo() 方法比较集合中对象的大小,然后进行升序排序,这种方式称为自然排序。

3. 客户化排序

java.util.Comparator 接口用于指定具体的排序方式,它有个 compare(Object obj1, Object obj2),用于比较两个对象的大小。

当表达式 compare(x,y) 的值大于 0,表示 x 大于 y;小于 0,表示 x 小于 y;等于 0,表示 x 等于 y。如果想让 TreeSet 按照 Customer 对象的 name 属性进行降序排列,可以先创建实现 Comparator 接口的类 CustomerComparator,例如,

```
import java.util.*;
public class CustomerComparator implements Comparator
{
    public int compare(Object o1, Object o2)
    {
        Customer c1 = (Custoemr)o1;
        Customer c2 = (Customer)o2;
        if(c1.getName().compareTo(c2.getName())>0) return -1;
        if(c1.getName().compareTo(c2.getName())<0) return 1;
        return 0;
    }
}
```

接下来在构造 TreeSet 的实例时,调用它的 TreeSet(Comparator comparator) 构造方法:

```
Set set = new TreeSet(new CustomerComparator());
Customer c1 = new Customer("TOM",15);
Customer c2 = new Customer("JACK",20);
Customer c3 = new Customer("MIKE",38);
set.add(c1); set.add(c2); set.add(c3);
Iterator it = set.iterator();
while(it.hasNext())
{ Custoemr customer = (Customer)it.next();
    System.out.println(customer.getName()+""+customer.getAge(););
}
```

当 TreeSet 向集合中加入 Customer 对象时,会调用 CustomerComparator 类的 compare() 方法进行排序,以上 Tree 按照 Custoemr 对象的 name 属性进行降序排列,最后输出为:
TOM 15   MIKE 38 JACK 16

二、List 有重复、有序

List(列表):对象以线性方式存储,集合中的对象按索引位置排序,可以有重复对象,允许按照对象在集合中的索引位置检索对象。

实现类有 LinkedList、ArrayList 和 Vector,LinkedList 采用链表数据结构,而 ArrayList 代表大小可变的数组,Vector 和 ArrayList 比较相似,两者的区别在于 Vecotr 类的实现采用了同步机

制,而 ArrayList 没有使用同步机制。

List 按索引排列:

```
List list = new ArrayList();
list.add(new Integer(3));
list.add(new Integer(4));
list.add(new Integer(3));
list.add(new Integer(2));
```

List 的 get(int index)方法返回集合中由参数 index 指定的索引位置的对象,第一个加入到集合中的对象的索引位置为 0:

```
for( int i=0,i<list.size;i++)
{
    System.out.println(list.get(i));
}
```

输出结果为:3 4 3 2

List 只能对集合中的对象按索引位置排序,如果希望对 List 中的对象按其他特定方式排序,可以借助 Comparator 接口和 Collections 类。

Collections 类是 Java 集合 API 中的辅助类,它提供了操纵集合的各种静态方法,其中 sort()方法用于对 List 中的对象进行排序:

sort(List list):对 List 中的对象进行自然排序。

sort(List list,Comparator comparator):对 List 中的对象进行客户化排序,comparator 参数指定排序方式。

对以下 List 进行自然排序:

```
List list = new ArrayList();
list.add(new Integer(3));
list.add(new Integer(4));
list.add(new Integer(3));
list.add(new Integer(2));
Collections.sort(list);
for(int i=0;i<list.size();i++)
{
    System.out.println(list.get(i));
}
```

以上输出结果:2 3 3 4

ArrayList(数组表)——有重复、有顺序。类似于 Vector,都用于缩放数组维护集合。区别如下:

(1)同步性:Vector 是线程安全的,也就是说是同步的,而 ArrayList 是线程序不安全的,不是同步的。

(2)数据增长:当需要增长时,Vector 默认增长为原来一培,而 ArrayList 却是原来的一半。

LinkedList(链表)是双向链表,适合变更很多的 List。

用在 FIFO,用 addList()加入元素 removeFirst()删除元素。

用在 FILO,用 addFirst()/removeLast()。

ListIterator 提供双向遍历 next() previous(),可删除、替换、增加元素。

(3) Map(映射):集合中的每一个元素包含一对键对象和一对值对象,集合中没有重复的键对象,值对象可以重复,它的有些实现类能对集合中的键对象进行排序:

Map map=new HashMap( );

map. put("1","Mon");

map. put("1",Monday);

map. put("2","monday");

由于第一次和第二次加入到 Map 中的键对象都是1,所以第一次加入的值对象将被覆盖,而第二个和第三个的值对象虽然相同,但是键对象不一样,所以分配了不同的地址空间,所以不会覆盖,也就是说一共有两个元素在 Map 集合中。

Map 有两种比较常用的实现:HashMap 和 TreeMap。Hashmap 按照哈希算法来存取键对象,有很好的存取能力,为了保证 HashMap 能正常工作,和 HashSet 一样,要求当两个键对象通过 equals( )方法比较为 true 时,这两个键对象的 hashCode( )方法返回的哈希码也一样。用于关键字/数值对,较高的存取性能。不允许重复的 key,但允许重复的 Value。

处理 Map 的三种集合:关键字集 KeySet( )、数值集 value( )、项目集 enrySet( )。四个具体版本如下:

①HashMap:散列表的通用映射表,无序,可在初始化时设定其大小,自动增长。

②LinkedHashMap:扩展 HashMap,对返回集合迭代时,维护插入顺序。

③WeakHashMap:基于弱引用散列表的映射表,如果不保持映射表外的关键字的引用,则内存回收程序会回收它。

④TreeMap:基于平衡树的映射表。TreeMap 实现了 SortedMap 接口,能对键对象进行排序,和 TreeSet 一样,TreeMap 也支持自然排序和客户化排序两种方式,以下程序中的 TreeMap 会对四个字符串类型的键对象"1"、"3"、"4"、"2"进行自然排序:

Map map=new TreeMap( );

map. put("1","Monday");

map. put("3","Wendsday");

map. put("4","Thursday");

map. put("2","Tuesday");

//返回集合中所有键对象的集合

Set keys=map. keySet( );

Iterator it=keys. iterator( );

while(it. hasNext)

{

  String key=(String)it. next( );

  //根据键对象得到值对象

  String value=(String)map. get(key);

  System. out. println(key+""+value);

}

以上输出结果为:

1 Monday 2 Wendsday 3 Thursday 4 Tuesday

如果希望 TreeMap 进行客户化排序,可以调用它的另一个构造方法 TreeMap(Comparator comparator),参数 comparator 指定具体的排序方式。

9.2 课后习题解答

一、选择题

1~3 CD ABC C

二、判断题

1~6 ××√√××

三、简答题

1. Vector 与 ArrayList 的区别是什么？

答：Vector 是线程同步的，所以它也是线程安全的，而 Arraylist 是线程异步的，是不安全的。如果不考虑到线程的安全因素，一般用 Arraylist 效率比较高。

如果集合中的元素的数目大于目前集合数组的长度时，vector 增长率为目前数组长度的 100%，而 Arraylist 增长率为目前数组长度的 50%。如果在集合中使用数据量比较大的数据，用 Vector 有一定的优势。

如果查找一个指定位置的数据，Vector 和 Arraylist 使用的时间是相同的，都是 0(1)，这个时候使用 Vector 和 Arraylist 都可以。而如果移动一个指定位置的数据花费的时间为 0(n-i) n 为总长度，这个时候就应该考虑到使用 linklist，因为它移动一个指定位置的数据所花费的时间为 0(1)，而查询一个指定位置的数据时花费的时间为 0(i)。

2. HashMap 与 TreeMap 的区别是什么？

答：HashMap 通过 hashcode 对其内容进行快速查找，而 TreeMap 中所有的元素都保持着某种固定的顺序，如果你需要得到一个有序的结果你就应该使用 TreeMap（HashMap 中元素的排列顺序是不固定的）。HashMap 中元素的排列顺序是不固定的。

HashMap 通过 hashcode 对其内容进行快速查找，而 TreeMap 中所有的元素都保持着某种固定的顺序，如果你需要得到一个有序的结果你就应该使用 TreeMap（HashMap 中元素的排列顺序是不固定的）。"集合框架"提供两种常规的 Map 实现：HashMap 和 TreeMap（TreeMap 实现 SortedMap 接口）。

在 Map 中插入、删除和定位元素，HashMap 是最好的选择。但如果要按自然顺序或自定义顺序遍历键，那么 TreeMap 会更好。使用 HashMap 要求添加的键类明确定义了 hashCode() 和 equals() 的实现。这个 TreeMap 没有调优选项，因为该树总处于平衡状态。

结过研究，还发现了一点，二树 map 一样，但顺序不一样，导致 hashCode() 不一样。

3. Set 里的元素是不能重复的，那么用什么方法来区分重复与否呢？是用==还是 equals()？它们有何区别？

答：Set 里的元素是不能重复的，用 iterator() 方法来区分重复与否。

equals() 是判读两个 Set 是否相等。

equals() 和==方法决定引用值是否指向同一对象 equals() 在类中被覆盖，为的是当两个分离的对象的内容和类型相配的话，返回真值。

四、编程题

某中学有若干学生(学生对象放在一个 List 中),每个学生有一个姓名属性、班级名称属性(String)和考试成绩属性(int),某次考试结束后,每个学生都获得了一个考试成绩。请打印出每个班级的总分和平均分。

```java
public class TestStudent2 {
    public static void main(String[] args) {
        List students = new ArrayList();
        students.add(new Student("Liucy","0701",100));
        students.add(new Student("Huxz","0702",150));
        students.add(new Student("George","0702",142));
        students.add(new Student("Wanglin","0701",80));
        students.add(new Student("Wuwl","0701",91));
        students.add(new Student("Wangr","0702",100));
        Map m = new HashMap();
        for(int i=0;i<students.size();i++){
            Student s = (Student)students.get(i);
            String classNumber = s.getClassNumber();
            if(m.containsKey(classNumber)){
                List list = (List)m.get(classNumber);
                list.add(s);
            }
            else{
                List list = new ArrayList();
                list.add(s);
                m.put(classNumber,list);
            }
        }
        Set keys = m.keySet();
        Iterator it = keys.iterator();
        while(it.hasNext()){
            Object classNumber = it.next();
            List s = (List)m.get(classNumber);
            int total = 0;
            for(int n=0;n<s.size();n++){
                total = total+((Integer)((Student)s.get(n)).getScore()).intValue();
            }
            System.out.println(classNumber+"班总分:"+total+" 平均分:"+total/s.size());
        }
    }
}
class Student{
    private String name;
```

```java
    private String classNumber;
    private int score;
    public Student(String name, String classNumber, int score) {
        super();
        this.name = name;
        this.classNumber = classNumber;
        this.score = score;
    }
    public String getClassNumber() {
        return classNumber;
    }
    public void setClassNumber(String classNumber) {
        this.classNumber = classNumber;
    }
    public String getName() {
        return name;
    }
    public void setName(String name) {
        this.name = name;
    }
    public int getScore() {
        return score;
    }
    public void setScore(int score) {
        this.score = score;
    }
}
```

9.3 上机实验

一、实验目的与意义

1. 掌握集合 set 设计及使用；
2. 掌握集合 list 设计及使用；
3. 掌握集合 map 设计及使用。

二、实验内容

1. 掌握 HashSet 对象的使用；
2. 掌握 ArrayList 对象的使用；
3. 掌握 HashMap 对象的使用；
4. 综合使用集合案例。

三、实验要求

1. JDK1.5 与 eclipse 开发工具；
2. 了解不同集合特点、迭代器的使用。

9.4 程序代码

1. 集合类 Set 的设计

```java
import java.util.HashSet;
import java.util.Iterator;
import java.util.Set;
public class HashSetTest {
    public static void main(String[] args) {
        //加入 HashSet 中的对象会自动调用 equals( )方法
        System.out.println("* * * * * * * * * HashSet Test * * * * * * * *");
        Set set = new HashSet( );
        Student stu = new Student("0001","jack","male",27);
        Student stu2 = new Student("0002","kevin","male",30);
        Student stu3 = new Student("0002","kevin","male",30);
        Student stu4 = new Student("0004","kevin2","male",30);
        Student stu5 = new Student("0004","kevin2","male",30);
        set.add(stu);
        set.add(stu2);
        set.add(stu3);
        set.add(stu4);
        set.add(stu5);
        outSet(set);
    }
    private static void outSet(Set set) {
        Iterator iter = set.iterator( );
        while(iter.hasNext( )) {
            System.out.println(iter.next( ));
        }
    }
}
```

2. 集合类 List 的设计

```java
import java.util.ArrayList;
import java.util.Iterator;
import java.util.LinkedList;
import java.util.List;
import java.util.Vector;
public class ListTest {
    public static void main(String[] args) {
```

```java
//加入ArrayList中的对象不会去调用或覆盖equals()方法
System.out.println("* * * * * * * * * ArrayList Test * * * * * * * * * * * *");
List list = new ArrayList();
Student s = new Student("0001","jack","male",27);
Student s2 = new Student("0002","kevin","male",30);
Student s3 = new Student("0003","marry","female",25);
Student s4 = new Student("0003","marry","female",25);
list.add(new Integer(1));
list.add(s);
list.add(new Integer(2));
list.add(s2);
list.add(new Integer(3));
list.add(s3);
list.add(new Integer(3));
list.add(s4);
outListByIterator(list);
System.out.println("-----------------------------------------------------");
outListByIndex(list);
System.out.println();
System.out.println("* * * * * * * * * * * Vector Test * * * * * * * * * * *");
list = new Vector();
s = new Student("0004","larry","male",40);
s2 = new Student("0005","geroge","male",35);
s3 = new Student("0005","geroge","male",35);
list.add(s);
list.add(s2);
list.add(s3);
outListByIterator(list);
System.out.println("-----------------------------------------------------");
outListByIndex(list);
System.out.println();
System.out.println("* * * * * * * * * * * LinkedList Test * * * * * * * * * * * *");
list = new LinkedList();
s = new Student("0006","tom","male",30);
s2 = new Student("0007","john","male",20);
list.add(new Integer(1));
list.add(s);
list.add(new Integer(2));
list.add(s2);
outListByIterator(list);
System.out.println("-----------------------------------------------------");
outListByIndex(list);
}
private static void outListByIterator(List list) {
```

```java
    Iterator iter=list.iterator();
    while(iter.hasNext()){
        System.out.println(iter.next());
    }
}
private static void outListByIndex(List list){
    for(int i=0;i < list.size();i++){
        System.out.println(list.get(i));
    }
}
}
```

3. 集合类 Map 的设计

```java
import java.util.HashMap;
import java.util.Hashtable;
import java.util.Iterator;
import java.util.Map;
import java.util.Set;
public class HashMapTest {
public static void main(String[] args){
    System.out.println("*************HashMap Test***************");
    Map map=new HashMap();
    Student stu=new Student("0001","jack","male",35);
    Student stu2=new Student("0002","kevin","male",30);
    Student stu3=new Student("0004","larry","male",40);
    Student stu4=new Student("0005","john","male",25);
    map.put(new Integer(1),stu);
    map.put(new Integer(1),stu4);
    map.put(new Integer(2),stu3);
    map.put(new Integer(2),stu4);
    map.put(new Integer(2),stu4);
    map.put(new Integer(3),stu3);
    map.put(new Integer(3),stu4);
    System.out.println("**********outMapByKey***********");
    outMapByKey(map);
    System.out.println("************outMapByEntry************");
    outMapByEntry(map);
    System.out.println("*** testing different HashMap and HashTable ***");
    HashMap hashMap=new HashMap();
    hashMap.put(null, null);
    Hashtable hashtable=new Hashtable();
    System.out.println("hashMap="+hashMap.toString());
}
private static void outMapByKey(Map map){
    Set set=map.keySet();
```

```java
    Iterator iter=set.iterator();
    while(iter.hasNext()){
        Object key=iter.next();
        Object value=map.get(key);
        System.out.println("key="+key+" value="+value);
    }
}
private static void outMapByEntry(Map map){
    Set set=map.entrySet();
    Iterator iter=set.iterator();
    while(iter.hasNext()){
        Map.Entry entry=(Map.Entry)iter.next();
    }
}
}
```

4. 综合业务设计步骤

(1)集合的输入,可以是任意的类型;
(2)计算并输出给定集合的交集、并集和差集;
(3)求两个字符串数组的并集,利用 set 的元素唯一性;
(4)求两个数组的交集;
(5)求两个数组的差集;
(6)找出较长的数组来减较短的数组。

```java
package string;
import java.util.HashMap;
import java.util.HashSet;
import java.util.LinkedList;
import java.util.Map;
import java.util.Map.Entry;
import java.util.Set;

public class StringArray {
    public static void main(String[] args) {
        //测试 union
        String[] arr1 = {"hbhc", "bhuyg", "nyhv", "abc"};
        String[] arr2 = {"jgfc", "mhjg", "hbhc", "d", "abc"};
        String[] result_union = union(arr1, arr2);
        System.out.println("求并集的结果如下:");
        for (String str : result_union) {
            System.out.println(str);
        }
        System.out.println("----------------------分割线----------------------");

        //测试 insect
```

```java
        String[] result_insect = intersect(arr1, arr2);
        System.out.println("求交集的结果如下:");
        for (String str : result_insect) {
            System.out.println(str);
        }

        System.out.println("---------------------分割线---------------------");

        //测试 minus
        String[] result_minus = minus(arr1, arr2);
        System.out.println("求差集的结果如下:");
        for (String str : result_minus) {
            System.out.println(str);
        }
    }

//求两个字符串数组的并集,利用 set 的元素唯一性
public static String[] union(String[] arr1, String[] arr2) {
    Set<String> set = new HashSet<String>();
    for (String str : arr1) {
        set.add(str);
    }
    for (String str : arr2) {
        set.add(str);
    }
    String[] result = {};
    return set.toArray(result);
}

//求两个数组的交集
public static String[] intersect(String[] arr1, String[] arr2) {
    Map<String, Boolean> map = new HashMap<String, Boolean>();
    LinkedList<String> list = new LinkedList<String>();
    for (String str : arr1) {
        if (!map.containsKey(str)) {
            map.put(str, Boolean.FALSE);
        }
    }
    for (String str : arr2) {
        if (map.containsKey(str)) {
            map.put(str, Boolean.TRUE);
        }
    }
```

```java
        for (Entry<String,Boolean> e ; map.entrySet()) {
            if (e.getValue().equals(Boolean.TRUE)) {
                list.add(e.getKey());
            }
        }

        String[] result = {};
        return list.toArray(result);
    }


    //求两个数组的差集
    public static String[] minus(String[] arr1, String[] arr2) {
        LinkedList<String> list = new LinkedList<String>();
        LinkedList<String> history = new LinkedList<String>();
        String[] longerArr = arr1;
        String[] shorterArr = arr2;
        //找出较长的数组来减较短的数组
        if (arr1.length > arr2.length) {
            longerArr = arr2;
            shorterArr = arr1;
        }
        for (String str : longerArr) {
            if (!list.contains(str)) {
                list.add(str);
            }
        }
        for (String str : shorterArr) {
            if (list.contains(str)) {
                history.add(str);
                list.remove(str);
            } else {
                if (!history.contains(str)) {
                    list.add(str);
                }
            }
        }

        String[] result = {};
        return list.toArray(result);
    }
}
```

运行结果如图9.1、图9.2、图9.3所示。

```
输出 - string2 (run)
run:
求并集的结果如下：
d
abc
jgfc
nyhv
mhjg
hbhc
bhuyg
```

图 9.1　输出结果一

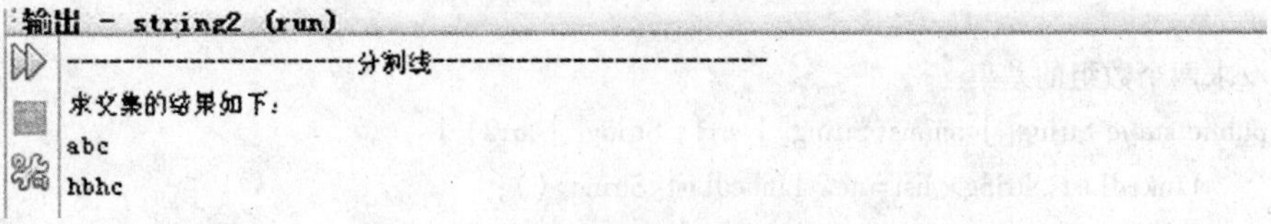

```
输出 - string2 (run)
------------------------分割线------------------------
求交集的结果如下：
abc
hbhc
```

图 9.2　输出结果二

```
输出 - string2 (run)
hbhc
------------------------分割线------------------------
求差集的结果如下：
bhuyg
nyhv
jgfc
mhjg
d
成功生成（总时间：1 秒）
```

图 9.3　输出结果三

第10章

多 线 程

10.1 典型例题解析

【例 10.1】 实现两个定时线程,其中一个线程每隔 1 秒显示一次,另一个线程每隔 3 秒显示一次。本例题采用从 Thread 类派生的方式创建了 Lefthand 和 Righthand 两个线程子类,在这两个子类中分别重载 run() 函数,在每个 run() 函数中先输出一个字符串,然后休眠一段时间。

```java
import java.io.*;
public class Time{
    static Lefthand left;         //声明静态的类的对象
    static Righthand right;
    public  static void main(String args[]){
        left=new Lefthand()    ;   //创建两个线程
        right=new Righthand();
        left.start();                 //通过 start() 方法启动线程
        right.start();
    }
}
class Lefthand extends Thread{        //lefthand 类继承 Thread 类
    public void run(){
        for(;;){
            System.out.println("每 1 秒显示一次!");
            try{
                sleep(1000);           //使当前活动线程睡眠 1 秒
            }
            catch(InterruptedException e){}   //捕捉异常
        }
    }
}
class Righthand extends Thread{
    public void run(){
```

```
        for(;;){
            System.out.println("每3秒显示一次!");
        Try{
            sleep(3000);    //使当前活动线程睡眠3秒
        }
        catch(InterruptedException e){}
        }
    }
}
```

程序运行结果如图 10.1 所示。

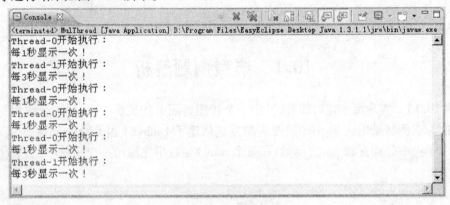

图 10.1　程序运行结果

【例 10.2】　本例题使用两个线程模拟用户同时从银行取钱。类 Cbank 模拟银行；其中静态变量 s 表示银行存款；静态方法 sub( )表示取款操作；参数 m 表示每次的取款额。每次取款的操作过程是首先将账户中的现有存款额 s 的值暂时存到临时变量 temp 中，再从 temp 中减去取款值 m。为了模拟银行取款过程中的网上阻塞，让系统休眠一随即时间段，最后显示最新存款额 s。

```
import java.io.*;
class CBank{
    private static int s=2000;
    public synchronized static void sub(int m){
        int temp=s;
        temp=temp-m;
        try{
            Thread.sleep((int)(1000*Math.random()));  //随即休眠一段时间
        }
        catch(InterruptedException e){
        }
        s=temp;     //取过钱后更新存款值 s
        System.out.println(Thread.currentThread().getName()+"在取款");
        System.out.println("存款余额为:"+s);    //显示存款值余额
    }
}
class Customer extends Thread{
```

```
public void run( ){
    for( int i=1;i<=3;i++) CBank.sub(100);   //每个用户取款三次
  }
}
public class ThreadSyn{
  public static void main(String[ ] args){
    Customer customer1=new Customer( );
    Customer customer2=new Customer( );
    customer1.start( );   //启动线程1
    customer2.start( );   //启动线程2
  }
}
```

程序运行结果如图 10.2 所示。

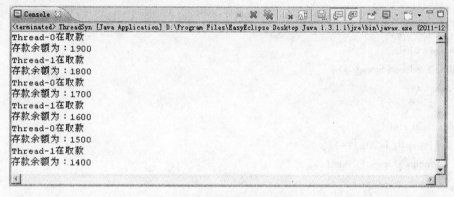

图 10.2　程序运行结果

【例 10.3】　本题是典型的生产者和消费者问题,生产者将生产出来的商品放入仓库,消费者将仓库中的商品拿走。关于生产者与消费者的原理请参考本章关于"多线程通信"部分的知识点说明。

```
import java.io.*;
//ProducerConsumer 是主类,Producer 生产者,Consumer 消费者,Product 产品
//Storage 仓库
public class ProducerConsumer {
  public static void main(String[ ] args) {
    Storage s=new Storage( );
    Producer p=new Producer(s);
    Consumer c=new Consumer(s);
    Thread tp=new Thread(p);
    Thread tc=new Thread(c);
    tp.start( );
    tc.start( );
  }
}
class Consumer implements Runnable {//消费者
  Storage s=null;
```

```java
    public Consumer(Storage s){
      this.s=s;
    }
    public void run(){
      for(int i=0; i<20; i++){
        Product p=s.pop();//取出产品
        try{
          Thread.sleep((int)(Math.random()*1500));
        }catch(InterruptedException e){
          e.printStackTrace();
        }
      }
    }
}
class Producer implements Runnable {//生产者
    Storage s=null;
    public Producer(Storage s){
      this.s=s;
    }
    public void run(){
      for(int i=0; i<20; i++){
        Product p=new Product(i);
        s.push(p);//放入产品
        System.out.println("生产者放入:"+p);
        try{
          Thread.sleep((int)(Math.random()*1500));
        }catch(InterruptedException e){
          e.printStackTrace();
        }
      }
    }
}
class Product {
    int id;
    public Product(int id){
      this.id=id;
    }
    public String toString(){//重写 toString 方法
      return "产品:"+this.id;
    }
}
class Storage {
    int index=0;
    Product[] products=new Product[5];
```

```
public synchronized void push(Product p){//放入产品
    while(index==this.products.length){
        try{
            this.wait();
        } catch (InterruptedException e){
            e.printStackTrace();
        }
    }
    this.products[index]=p;
    System.out.println("生产者放入"+index+"位置:"+p);
    index++;
    this.notifyAll();
}

public synchronized Product pop(){//取出
    while(this.index==0){
        try{
            this.wait();
        } catch (InterruptedException e){
            e.printStackTrace();
        }
    }
    index--;
    this.notifyAll();
    System.out.println("消费者从"+ index+ "位置取出:"+this.products[index]);
    return this.products[index];
}
}
```

程序运行结果如图 10.3 所示。

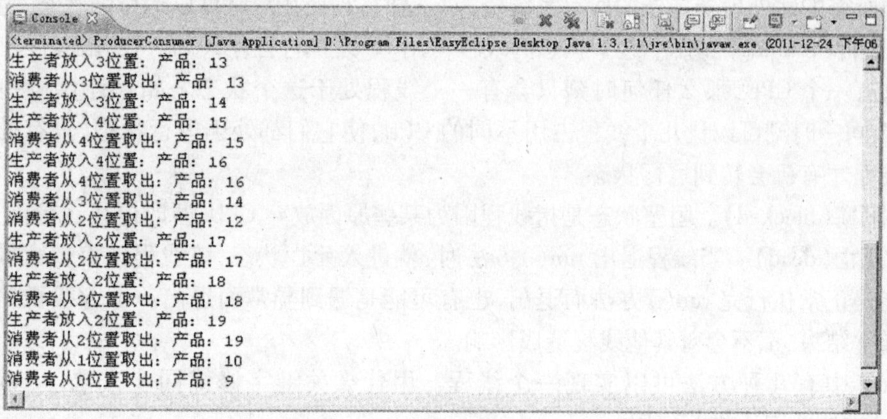

图 10.3 程序运行结果

10.2 课后习题解答

一、选择题

1～5 DDBAA   6～10 BABBD   11～15 DADAC   16～20 DDACC

二、填空题

1. java.lang.Thread   2. 并发   3. 抢先调度   4. stop()   5. 死亡   6. Thread   7. 可运行状态   8. 线程体   9. 返回线程的字符串信息   10. 用户   11. 寄存器   12. 虚拟的 CPU   代码 数据   13. MAX_PRIORITY   MIN_PRIORITY   14. getPrority()   setPrority()   15. syschronized   16. 可运行状态   阻塞状态   17. 封锁   18. 代码 一组寄存器   19. 相互独立   20. wait()方法

三、判断题

1～5 ×√××√   6～10 √×√√√
11～15 ×√√√×   16～20 ×√×√×

四、简答题

1. 线程的基本概念、线程的基本状态以及状态之间的关系。

答:线程指在程序执行过程中,能够执行程序代码的一个执行单位,每个程序至少有一个线程,也就是程序本身。

Java 中的线程有五种状态分别是:

(1) 新建(new)。用 new 语句创建的线程对处于新建状态,此时它和其他 Java 对象一样,仅仅在 Heap 中被分配了内存。当一个线程处于新建状态时,它仅仅是一个空的线程对象,系统不为它分配资源。Thread t = new Thread(new Runner());

(2) 就绪(runnable)。程序通过线程对象调用启动方法 start()后,系统会为这个线程分配它运行时所需的除处理器之外的所有系统资源。这时,它处在随时可以运行的状态。

(3) 运行(running)。处于这个状态的线程占用 CPU,执行程序代码。在并发环境中,如果计算机只有一个 CPU,那么任何时刻只会有一个线程处于这个状态。如果计算机中有多个CPU,那么同一时刻可以让几个线程占用不同的 CPU,使它们都处于运行状态,只有处于就绪状态的线程才有机会转到运行状态。

(4) 阻塞(blocked)。阻塞状态是指线程因为某些原因放弃 CPU,暂时停止运行。

(5) 死亡(dead)。当线程退出 run()方法时,就进入死亡状态,该线程结束生命周期。线程有可能是正常执行完 run()方法而退出,也有可能是遇到异常而退出。不管线程是正常结束还是异常结束,都不会对其他线程造成影响。

2. Java 中有几种方法可以实现一个线程?用什么关键字修饰同步方法？stop()和 suspend()方法为何不推荐使用?

答:有两种实现方法,分别是继承 Thread 类与实现 Runnable 接口。

用 synchronized 关键字修饰同步方法。不推荐使用 stop(),因为它不安全。它会解除由线程获取的所有锁定,而且如果对象处于一种不连贯状态,那么其他线程能在那种状态下检查和修改它们。结果很难检查出真正的问题所在。suspend( )方法容易发生死锁。调用 suspend( )的时候,目标线程会停下来,但却仍然持有在这之前获得的锁定。此时,其他任何线程都不能访问锁定的资源,除非被"挂起"的线程恢复运行。对任何线程来说,如果它们想恢复目标线程,同时又试图使用任何一个锁定的资源,就会造成死锁。所以不应该使用 suspend( ),而应在自己的 Thread 类中置入一个标志,指出线程应该活动还是挂起。若标志指出线程应该挂起,便用 wait( )命其进入等待状态。若标志指出线程应当恢复,则用一个 notify( )重新启动线程。

3. sleep( )和 wait( )有什么区别?

答:sleep 是线程类(Thread)的方法,导致此线程暂停执行指定时间,执行机会给其他线程,但是监控状态依然保持,到时会自动恢复。调用 sleep 不会释放对象锁。

wait 是 Object 类的方法,对一个对象调用 wait 方法导致本线程放弃对象锁,进入等待对象的等待锁定池,只有针对此对象发出 notify 方法(或 notifyAll)后本线程才进入对象锁定池准备获得对象锁进入运行状态。

4. 请说出你所知道的线程同步的方法。

答:(1) 同步代码块 synchronized(object){代码段}

(2)同步函数 public synchronized void sale( ){ //...}

5. 同步和异步有何异同?在什么情况下分别使用他们?举例说明。

答:如果数据将在线程间共享,例如,正在写的数据以后可能被另一个线程读到,或者正在读的数据可能已经被另一个线程写过了,那么这些数据就是共享数据,必须进行同步存取。

当应用程序在对象上调用了一个需要花费很长时间来执行的方法,并且不希望让程序等待方法的返回时,就应该使用异步编程,在很多情况下采用异步途径往往更有效率。

6. 当一个线程进入一个对象的一个 synchronized 方法后,其他线程是否可进?为什么?

答:(1)当一个线程进入一个对象的一个 synchronized 方法后,其他线程访问该对象的非同步方法。

(2)当一个线程进入一个对象的一个 synchronized 方法后,其他线程也访问该同步方法。

(3)当一个线程进入一个对象的一个 synchronized 方法后,其他线程同时访问该对象的另一个同步方法。

五、编程题

1. 编写一个多线程类,该类的构造方法调用 Thread 类带字符串参数的构造方法。建立自己的线程名,然后随机生成一段休眠时间,再将自己的线程名和休眠时间显示出来。该线程运行后,休眠一段时间,该时间就是在构造方法中生成的时间。最后编写一个测试类,创建多个不同名字的线程,并测试其运行情况。

```
public class MyThread extends Thread{
    int sleepTime;
    public MyThread(String s){
        super(s);
        sleepTime=(int)(500*Math.random());
        System.out.println("Name :"+getName()+"\tSleep:"+sleepTime);
```

```java
    }
    public void run(){
        try{
            sleep(sleepTime);
        }catch(Exception e){}
        System.out.println("Thread "+getName());
    }
}
public class TestThreads{
    public static void main(String[] args){
        MyThread thread0,thread1,thread2,thread3;
        thread0 = new MyThread("0");
        thread1 = new MyThread("1");
        thread2 = new MyThread("2");
        thread3 = new MyThread("3");
        thread0.start();
        thread1.start();
        thread2.start();
        thread3.start();
        try{
            System.in.read();
        }catch(Exception e){}
    }
}
```

2. 编写一个程序,测试异常。该类提供一个输入整数的方法,使用这个方法先输入两个整数,再用第一个整数除以第二个整数,当第二个整数为 0 时,抛出异常,此时程序要捕获异常。

```java
import java.io.*;
public class TestExceptions{
    static int getInt() throws IOException{
        BufferedReader input = new BufferedReader(new InputStreamReader(System.in));
        System.out.print("Enter an integer:");
        String s = input.readLine();
        return Integer.parseInt(s);
    }
    public static void main(String[] args){
        int n1=0,n2=0,n3=0;
        try{
            n1 = getInt();
            n2 = getInt();
            n3 = n1/n2;
        }catch(Exception e){
            System.out.println("["+e+"]");
        }
        System.out.println(n1+"/"+n2+"="+n3);
```

}
}

3. 编写一个用线程实现一个数字时钟的应用程序。该线程类要采用休眠的方式,把绝大部分时间让系统使用。

```
import java.awt.*;
import java.util.Calendar;
public class Clock extends Frame implements Runnable{
  Thread thread;
  Font font = new Font("Monospaced",Font.BOLD,18);
  int hour,minute,second;
  Clock(){
    if(thread = = null){
      thread = new Thread(this);
      thread.start();
    }
    setSize(200,100);
    setVisible(true);
  }
  public void run(){
    for(;;){
      Calendar time = Calendar.getInstance();
      hour = time.get(Calendar.HOUR);
      minute = time.get(Calendar.MINUTE);
      second = time.get(Calendar.SECOND);
      try{
         thread.sleep(1000);
      }catch(Exception e){}
      repaint(1000);
    }
  }
  public void paint(Graphics g){
    g.setFont(font);
  String time = String.valueOf(hour)+":"+String.valueOf(minute)+":"+
               String.valueOf(second);
    g.drawString(time, 50, 50);
  }
  public static void main(String[] args){
    Clock c = new Clock();
  }
}
```

4. 编写一个使用继承 Thread 类的方法实现多线程的程序。该类有两个属性:一个字符串代表线程名,一个整数代表该线程要休眠的时间。线程执行时,显示线程名和休眠时间。

```
public class ThreadDemo extends Thread{
```

```java
    private String whoAmI;
    private int delay;
    public ThreadDemo(String s,int d){
      whoAmI=s;
      delay=d;
    }
    public void run(){
      try{
        sleep(delay);
      }catch(InterruptedException e){}
      System.out.println("Hello! I am "+whoAmI+" I slept "+delay+" milliseconds");
    }
}
```

5. 应用继承类 Thread 的方法实现多线程类,该线程 3 次休眠若干(随机)毫秒后显示线程名和第几次执行。

```java
public class MyThread1 extends Thread{
    public void run(){
      try{
        for(int i=0;i<3;i++){
          int msec=(int)(1000*Math.random());
          Thread.sleep(msec);
          System.out.println(getName()+"的第"+i+"次执行");
        }
      }catch(InterruptedException ex){
        ex.printStackTrace();
      }
    }
}
```

10.3 上 机 实 验

一、实验目的与意义

1. 理解线程的概念;
2. 掌握通过继承 Thread 类创建线程的方法;
3. 掌握通过使用 Runnable 接口创建线程的方法;
4. 理解多线程同步的概念,并掌握多线程同步的方法。

二、实验内容

1. 编写一个 java 应用程序,实现汉字打字练习功能,掌握通过继承 Thread 类创建线程。
2. 使用 Runnable 接口创建两个线程。选择两个城市作为预选旅游目标,实现两个独立的线程分别显示 10 次城市名,每次显示后休眠一段随机时间(1 000 毫秒以内),哪个先显示完毕，就决定去哪个城市。

3. 编程实现一个多线程的火车票预订系统，除了正常售票外，还要保障售出的车票不被再次出售。重点练习和掌握实现线程同步的两种方法：同步代码块和同步方法。

三、实验要求

1. JDK1.5 与 eclipse 开发工具；
2. 掌握 Thread 抽象类和 Runnable 接口的编程。

10.4　程序代码

1. 汉字打字练习程序

在主线程中创建一个 Frame 类型窗口，并在该窗口中创建 1 个名为 giveWord 的线程。线程 Word 每隔 2 秒给出一个汉字，用户在文本框中输入所给的汉字，若正确，得分加 1。

```
import java.awt.*;
import java.awt.event.*;
public class MainClass{
public static void main(String[] args){
    // TODO Auto-generated method stub
    new MainFrame();
}
}
class WordThread extends Thread{
//通过继承 Thread 类创建实现线程类
char word;
int k=19968;
Label show;
WordThread(Label show){
    this.show=show;
}
public void run(){
//重载 run 函数
    k=19968;
    while(true){
        word=(char)k;
        show.setText(""+word);
        try{
            sleep(6000);
        }
        catch(InterruptedException e){}
        k++;
        if(k>=29968) k=19968;
    }
}
```

```java
}
class MainFrame extends Frame implements ActionListener{
Label wordLabel;
Button button;
TextField inputText,scoreText;
WordThread giveWord;
int score=0;
MainFrame(){
   wordLabel=new Label(" ",Label.CENTER);
   wordLabel.setFont(new Font("",Font.BOLD,72));
   button=new Button("开始");
   inputText=new TextField(3);
   scoreText=new TextField(5);
   scoreText.setEditable(false);
   giveWord=new WordThread(wordLabel);
   button.addActionListener(this);
   inputText.addActionListener(this);
   add(button,BorderLayout.NORTH);
   add(wordLabel,BorderLayout.CENTER);
   Panel southPanel=new Panel();
   southPanel.add(new Label("输入标签所显示的汉字,然后回车:"));
   southPanel.add(inputText);
   southPanel.add(scoreText);
   add(southPanel,BorderLayout.SOUTH);
   setBounds(100,100,350,180);
   setVisible(true);
   validate();
   addWindowListener(new WindowAdapter(){
       public void windowClosing(WindowEvent e){
           System.exit(0);
       }
   });
}
public void actionPerformed(ActionEvent e){
   if(e.getSource()==button){
     if(!(giveWord.isAlive())){
        giveWord=new WordThread(wordLabel);
     }
     try{
         giveWord.start();
     }
     catch(Exception exe){}
   }
   else if(e.getSource()==inputText){
```

```
        if(inputText.getText().equals(wordLabel.getText())){
            score++;
        }
        scoreText.setText("得分"+score);
        inputText.setText(null);
    }}
}
```

程序运行结果如图 10.4 所示。

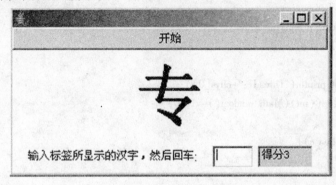

图 10.4　程序运行结果

2. 选择旅游城市

选择两个城市作为预选旅游目标,实现两个独立的线程分别显示 10 次城市名,每次显示后休眠一段随机时间(1 000 毫秒以内),哪个先显示完毕,就决定去哪个城市。请用 Runnable 接口创建线程。

```
import java.awt.*;
import java.awt.event.*;
public class SelCity{
public static void main(String[] args){
    // TODO Auto-generated method stub
    String[] citys={"北京","上海"};
    ThreadCity1 thread1=new ThreadCity1(citys);
    ThreadCity2 thread2=new ThreadCity2(citys);
    Thread th1=new Thread(thread1);
    Thread th2=new Thread(thread2);
    th1.start();
    th2.start();
    while(th1.isAlive()||th2.isAlive()){}
    if(thread1.getTime()<thread2.getTime()){
      System.out.println("决定去:"+citys[0]);
    }
    else{
      System.out.println("决定去:"+citys[1]);
    }}
}
class ThreadCity1 implements Runnable{
```

```java
    private String[ ] citys;
    private long time;
    public long getTime( ) {
        return time;
    }
    ThreadCity1(String[ ] citys) {
        this.citys = citys;
    }
    public void run( ) {
        long start = System.currentTimeMillis( );
        for(int i = 0; i<10; i++) {
            try {
                System.out.println("Thread1:"+citys[0]);
                Thread.sleep((int)(Math.random( ) * 1000));
            }
            catch(Exception e) {
            }
        }
        time = System.currentTimeMillis( ) - start;
        System.out.println("Thread1 Ends!");
    }
}
class ThreadCity2 implements Runnable {
    private String[ ] citys;
    private long time;
    ThreadCity2(String[ ] citys) {
        this.citys = citys;
    }
    public long getTime( ) {
        return time;
    }
    public void run( ) {
        long start = System.currentTimeMillis( );
        for(int i = 0; i<10; i++) {
            try {
                System.out.println("Thread2:"+citys[1]);
                Thread.sleep((int)(Math.random( ) * 1000));
            }
            catch(Exception e) {
            }
        }
        time = System.currentTimeMillis( ) - start;
        System.out.println("Thread2 Ends!");
    }
}
```

}

程序运行结果如图 10.5 所示。

```
<terminated> SelCity [Java Application] D:\Program Files\EasyEclipse Desktop Java 1.3.1.1\jre\bin\javaw
Thread1:北京
Thread2:上海
Thread1:北京
Thread2:上海
Thread1:北京
Thread1:北京
Thread2 Ends!
Thread1:北京
Thread1:北京
Thread1:北京
Thread1:北京
Thread1 Ends!
决定去:上海
```

图 10.5 程序运行结果

3. 火车订票系统

一个多线程的火车票预订系统如果在车票被售出后没有及时更新数据库中的信息会导致在同一时刻该火车票被另一个乘客购买,这一问题被称为线程安全问题,为了解决此问题,必须引入进程同步机制。实现线程同步有两种方法:分别是同步代码块和同步方法。在本实验中分别使用这两种方法来实现线程的同步,确保售出的车票不会再次被售出。

(1)使用同步代码块实现进程同步:

```java
public class ticket {
public static void main(String[] args) {
    // TODO Auto-generated method stub
    SaleTickets m = new SaleTickets();
    Thread t1 = new Thread(m,"System 1");
    Thread t2 = new Thread(m,"System 2");
    t1.start();
    t2.start();
}}
class SaleTickets implements Runnable{
private String ticketNo = "20111201";
private int ticket = 1;
public void run() {
    System.out.println(Thread.currentThread().getName()+" is sailing Ticket "+ticketNo);
    synchronized(this) {
        if(ticket>0) {
            try {
                Thread.sleep((int)(Math.random()*1000));
            } catch(InterruptedException e) {}
            ticket = ticket-1;
            System.out.println("ticket is saled by "+Thread.currentThread().getName()+",
                amount is: "+ticket);
```

```
        }
        else
            System.out.println("Sorry "+Thread.currentThread().getName()+",Ticket
                "+ticketNo+" is saled");
    }}
}
```

(2) 使用同步方法块实现进程同步：

```java
public class syn2 {
    public static void main(String[] args) {
        // TODO Auto-generated method stub
        SaleTickets m=new SaleTickets();
        Thread t1=new Thread(m,"System 1");
        Thread t2=new Thread(m,"System 2");
        t1.start();
        t2.start();
    }
}
class SaleTickets implements Runnable{
    private String ticketNo="20111201";
    private int ticket=1;
    public void run(){
        System.out.println(Thread.currentThread().getName()+" is sailing Ticket "+ticketNo);
        sale();
    }
    public synchronized void sale(){
        if(ticket>0){
            try{
                Thread.sleep((int)(Math.random()*1000));
            }catch(InterruptedException e){
                e.printStackTrace();
            }
            ticket=ticket-1;
            System.out.println("ticket is saled by "+Thread.currentThread().getName()+",
                amount is: "+ticket);
        }
        else
            System.out.println("Sorry "+Thread.currentThread().getName()+",Ticket "+ticketNo+"
                is saled");
    }
}
```

程序运行结果如图 10.6 所示。

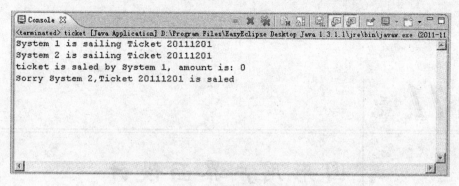

图 10.6　程序运行结果

第11章

图形用户界面设计

11.1 典型例题解析

【例 11.1】 编写一个程序,在屏幕上显示带标题的窗口,并添加一个按钮,当用户单击按钮时,结束程序。

解析:具有事件处理功能的 GUI 程序需要导入 java.awt 和 java.awt.event 包中的所有类,并定义一个继承 Frame 的类,定义组件,为组件注册事件监听器,实现或覆盖监听器接口或适配器类中的方法,进行事件处理。

本例中,定义了 Button 组件 quit,quit 的监听器是实现 ActionListener 接口的 FirstFrame 类对象 this。FirstFrame 实现了 ActionListener 接口中的 actionPerformed( )方法,其功能是结束程序运行。

```
import java.awt.*;
import java.awt.event.*;
public class FirstFrame extends Frame implements ActionListener{
private Button quit=new Button("退出");
public FirstFrame()
{
   super("First window");
   add(quit);
   quit.addActionListener(this);
   setSize(250,80);
   setVisible(true);
}
public void actionPerformed(ActionEvent e){
   System.exit(0);
}
public static void main(String args[]){
   FirstFrame ff=new FirstFrame();
}
}
```

程序运行界面如图 11.1 所示。如果单击"退出"按钮,关闭窗口,结束程序运行。

图 11.1 FirstFrame 运行界面

【例 11.2】 编写一个窗口程序,窗口中有两个按钮和一个文本行。当单击第一个按钮时,结束程序运行;当单击第二个按钮时,文本行显示该按钮被单击的次数。

解析:定义了 Button 组件 quit 和 click 及 TextField 组件 txf。quit 和 click 的监听器都是实现 ActionListener 接口的 ClickButton 类对象 this。ClickButton 实现了 ActionListener 接口中的 actionPerformed()方法。在 actionPerformed()中,通过 getSource()方法判断引起 ActionEvent 事件的事件源。如果事件源是 quit,结束程序运行;如果事件源是 click,首先给 count 变量的值加 1,再在 txf 文本行中显示 count 值,即 click 被单击的次数。

```
import java.awt.*;
import java.awt.event.*;
public class ClickButton extends Frame implements ActionListener{
    private Button quit=new Button("退出");
    private Button click=new Button("单击");
    private TextField txf=new TextField("还未单击按钮");
    private int count=0;
    public ClickButton(){
        super("Click Button");
        setLayout(new FlowLayout());
        add(quit);
        add(click);
        add(txf);
        quit.addActionListener(this);
        click.addActionListener(this);
        setSize(250,100);
        setVisible(true);
    }
    public void actionPerformed(ActionEvent e)
    {
        if(e.getSource()==quit)
            System.exit(0);
        else if(e.getSource()==click)
        {
            count++;
            txf.setText("单击了"+Integer.toString(count)+"次");
        }
    }
    public static void main(String args[]){
        ClickButton cb=new ClickButton();
```

运行结果如图 11.2 所示。

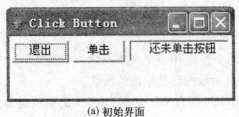

(a) 初始界面

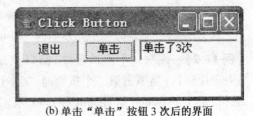

(b) 单击"单击"按钮 3 次后的界面

图 11.2 运行结果

【例 11.3】 编写一窗口程序,实现以下功能:

当用户按下 q 键或单击窗口关闭按钮时,结束程序运行;当用户按其他字符时,在屏幕上显示被按下的字符。

解析:要对键盘按键响应,需要对 KeyEvent 事件监听。KeyEvent 事件监听器是实现 KeyListener 接口的类对象,该类要实现 KeyListener 接口中的抽象方法 keyPressed( ),keyReleased( )和 keyTyped( )。

要响应关闭窗口操作,需要对 WindowEvent 事件监听。WindowEvent 事件监听器是实现 WindowListener 接口的类对象,该类要实现 WindowListener 接口中的抽象方法 windowClosing,windowClosed( ),windowDeactivated( ),windowActived( ),windowIconified( ),windowDeiconified( )和 windowOpened( )。

本例中,内部 KeyHandler 实现 KeyListener 接口,并实现其中的抽象方法 keyPressed( ),keyReleased( )和 keyTyped( )。KeyHandler 的对象担任 KeyEvent 事件的监听器。外部类 WindowHandler 实现 WindowListener 接口,并实现其中的 7 个抽象方法。WindowHandler 的对象 handler 担任 WindowEvent 事件的监听器。

监听器所属类可以定义为内部类(如 KeyHandler)、外部类(如 WindowHandler)或匿名类,其效果相同。监听器所属类不需要构造方法。

```java
import java.awt.*;
import java.awt.event.*;
public class MultiEvent extends Frame {
    private WindowHandler handler = new WindowHandler();
    public MultiEvent() {
        super("MutiEvent");
        setLayout(new FlowLayout());
        addKeyListener(new KeyHandler());
        addWindowListener(handler);
        setSize(250, 100);
        setVisible(true);
    }
    public static void main(String args[]) {
        MultiEvent me = new MultiEvent();
    }
```

```java
class KeyHandler implements KeyListener {
    public void keyPressed(KeyEvent e) {
        if (e.getKeyChar() == 'q')
            System.exit(0);
    }
    public void keyReleased(KeyEvent e) {
    }
    public void keyType(KeyEvent e) {
        System.out.println(e.getKeyChar() + "is pressed!");
    }
}

class WindowHandler implements WindowListener {
    public void windowClosing(WindowEvent e) {
        System.exit(0);
    }
    public void windowClosed(WindowEvent e) {
    }
    public void windowActivated(WindowEvent e) {
    }
    public void windowDeactivated(WindowEvent e) {
    }
    public void windowIconified(WindowEvent e) {
    }
    public void windowDeiconified(WindowEvent e) {
    }
    public void windowOpened(WindowEvent e) {
    }
}
```

【例 11.4】 编写一个窗口程序,包含一个文本行、两个按钮和一个文本区。当用户单击第一个按钮时,将文本行中的文本添加到文本区;当用户单击第二个按钮时,添加一个换行符后,将文本行中的文本添加到文本区。

解析:要对第一个按钮 add 和第二个按钮 addln 的单击操作(触发 ActionEvent 事件)进行响应。要对其注册 ActionEvent 事件监听器。事件监听器由实现 ActionListener 接口的 Texts 类对象 this 担任。Texts 类实现了 ActionListener 接口中的 actionPerformed() 方法。在 actionPerformed() 方法中,通过 getSource 获得事件源。如果事件源是 add,仅将文本行 txf 中的文本添加到文本区 txa 中;如果事件源是 addln,首先在 txa 中添加一换行符,再将 txf 中的文本添加到 txa 中。

要对窗口关闭操作进行响应,需要对其注册 WindowEvent 事件监听器。事件监听器由继承 WindowAdapter 适配器类的内部类 WindowHandler 的对象担任。WindowHandler 覆盖了 WindowAdapter 中的 windowClosing() 方法,用户单击窗口关闭按钮时,结束程序运行。

为了使窗口内组件排列整齐、美观,程序中使用了面板 Panel 组件 pal。类似 Frame,Panel 也是一种容器类,可以包含 Button,TextField 和 List 等组件,可以设置版面。将其他组件放入

Panel 组件后,再将 Panel 组件嵌入窗口,易于实现窗口内组件的合理布局。

```java
import java.awt.*;
import java.awt.event.*;
public class Texts extends Frame implements ActionListener{
    private TextField txf=new TextField();
    private TextArea txa=new TextArea();
    private Button add=new Button("Add");
    private Button addln=new Button("Addln");
    public Texts()
    {
        super("Texts");
        Panel pal=new Panel();
        pal.setLayout(new BorderLayout());
        pal.add(add,BorderLayout.WEST);
        pal.add(txf,BorderLayout.CENTER);
        pal.add(addln,BorderLayout.EAST);
        setLayout(new BorderLayout());
        add(pal,BorderLayout.NORTH);
        add(txa,BorderLayout.CENTER);
        add.addActionListener(this);
        addln.addActionListener(this);
        addWindowListener(new WindowHandler());
        setSize(300,200);
        setVisible(true);

    }
    public static void main(String args[]){
        Texts me=new Texts();
    }
    public void actionPerformed(ActionEvent e){
        if(e.getSource()==add)
            txa.append(txf.getText());
        else if(e.getSource()==addln)
            txa.append("\n"+txf.getText());
    }
}
class WindowHandler extends WindowAdapter{
    public void windowClosing(WindowEvent e){
        System.exit(0);
    }
}
```

运行结果如图 11.3 所示。

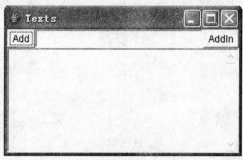 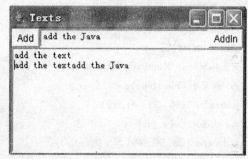

(a) 初始界面　　　　　　　　　　(b) 添加文本后的界面

图 11.3　运行结果

【例 11.5】　编程实现:在窗口中用不同颜色绘制椭圆、圆、圆弧和文本。

解析:继承 Frame 的类,图形绘制在 paint( )方法中进行。

通过"drawOval(20,40,30,50)"绘制椭圆,椭圆中心位于(20,40),短径(水平方向)和长径(垂直方向)分别为 30 和 50;通过"drawOval(85,35,90,90)"绘制圆,圆心位于(85,35),半径为 90;通过"drawArc(20,90,40,50,0,-150)"绘制圆弧,圆弧中心位于(20,90),短径(水平方向)和长径(垂直方向)分别为 40 和 50,圆弧角从 0 到 -150(顺时针为负);通过"filArc(20,160,40,50,0,180)"绘制填充圆弧,圆弧中心位于(20,160),短径和长径分别为 40 和 50,圆弧角从 0 到 180(逆时针为正);通过"drawstring("Painting",100,80)"从(100,80)点开始,显示字符串 Painting。图形颜色通过 setColor( )来设置。

要对窗口关闭操作进行响应,需要对其注册 WindowEvent 事件监听器。事件监听器由继承 WindowAdapter 适配器类的匿名类对象担任。匿名类覆盖了 WindowAdapter 中的 windowClosing( )方法,用户单击关闭按钮时,结束程序运行。

```
import java.awt.*;
import java.awt.event.*;
public class Drawing extends Frame{
public Drawing( ){
  super("Drawing");
  addWindowListener( new WindowAdapter( ){
    public void windowClosing(WindowEvent e)
    {
      System.exit(0);
    }
  });
  setSize(200,200);
  setVisible(true);
}
public static void main(String args[ ]){
  Drawing dr=new Drawing( );
}
public void paint(Graphics g){
  g.setColor(Color.red);
  g.drawOval(20, 40, 30, 50);
```

```
g.setColor(Color.green);
Font fnt=new Font("dialog",Font.ITALIC+Font.BOLD,15);
g.setFont(fnt);
g.drawString("Painting", 100, 80);
g.setColor(Color.blue);
g.drawOval(85, 35, 90, 90);
g.setColor(Color.pink);
g.drawArc(20, 90, 40, 50, 0, -150);
g.setColor(Color.black);
g.fillArc(20, 160, 40, 50, 0, 180);
    }
}
```

运行结果如图 11.4 所示。

图 11.4　Drawing 运行界面

11.2　课后习题解答

一、简答题

1. 简述 AWT 提供的基于事件监听模型的事件处理机制。

答:基于事件监听模型的事件处理是从一个事件源授权到一个或多个事件监听者,组件作为事件源可以触发事件,通过 addXXXlistener() 方法向组件注册监听器,一个组件可以注册多个监听器,如果组件触发了相应类型的事件,此事件被传送给已注册的监听器,事件监听器通过调用相应的实现方法来负责处理事件的过程。

2. 列出几个你所熟悉的 AWT 事件类,并举例说明什么时候会触发这些事件。

答:(1) ActionEvent 类:可以是鼠标单击按钮或者菜单,也可以是列表框的某选项被双击或文本框中的回车行为。

(2) KeyEvent 类:当用户按下或释放键时产生该类事件,也称为键盘事件。

(3) MouseEvent 类:当用户按下鼠标、释放鼠标或移动鼠标时会产生鼠标事件。

3. AWT 规定的 MouseEvent 类对应哪些监听器接口? 这些接口中都定义有哪些抽象方

法？

答：(1) MouseListener

public abstract void mouseClicked(MouseEvent mouseevent);

public abstract void mousePressed(MouseEvent mouseevent);

public abstract void mouseReleased(MouseEvent mouseevent);

public abstract void mouseEntered(MouseEvent mouseevent);

public abstract void mouseExited(MouseEvent mouseevent);

(2) MouseMotionListener

public abstract void mouseDragged(MouseEvent mouseevent);

public abstract void mouseMoved(MouseEvent mouseevent);

4. 简述 AWT 为何要给事件提供相应的适配器(即 Adapter 类)？

答：Java 规定：实现一个接口时必须对该接口的所有抽象方法进行具体的实现，哪怕有些抽象方法事件用户根本用不上，也要将其实现，比如上例中的 keyPressed( )方法，为此，Java 提供了一种叫作 Adapter(适配器)的抽象类来简化事件处理程序的编写。适配器类很简单，它其实就是一个实现了接口中所有抽象方法的"空"类，本身不提供实际功能。

5. 简述事件处理机制。

答：从 JDK1.1 之后，Java 采用委托事件模型。当事件发生时，事件源将事件对象传递给事件监听器处理。

事件处理基本过程为：

(1) 在程序中，向事件源注册事件监听器。

(2) 程序运行过程中，用户在事件源上引发某种事件时，Java 产生事件对象。

(3) 事件源将事件对象传递给事件监听器。

(4) 事件监听器根据事件对象的种类，调用相应的事件处理方法进行了事件处理。

二、程序设计

1. 设计程序实现：一个窗口包含文本行和标签，在文本行中输入一段文字并按回车键后，这段文字将显示在标签上。

解析：用内部类 ActionListener1 的对象监听文本行 text 上的 ActionEvent 事件，所以 ActionListener1 要实现 ActionListener 接口，实现其中的 actionPerformed( )方法。当在 text 中输入完文本，并按回车键后，执行 actionPerformed( )方法，通过 label.setText(text.getText( ))使 text 中的文本显示在标签 label 中。

用内部类 WindowAdapter1 的对象监听窗口的 WindowEvent 事件，所以 WindowAdapter1 要继承 WindowAdapter 类，覆盖其中的 windowClosing( )方法。当关闭窗口时，执行 WindowClosing( )方法，结束程序。

```
import java.awt.*;
import java.awt.event.*;
public class Show Text1 extends Frame{
    private TextField text=new TextField();
    private Label label=new Label("请输入文字:");
    public ShowText1()
```

```
        }
        super("Show Text");
        setLayout(new BorderLayout());
        label.setAlignment(Label.CENTER);
        add(text,BorderLayout.NORTH);
        add(label,BorderLayout.CENTER);
        text.addActionListener(new ActionListener1());
        addWindowListener(new WindowAdapter1());
        setSize(200,150);
        setVisible(true);
    }
    class ActionListener1 implements ActionListener{
        public void actionPerformed(ActionEvent e)
        {
            label.setText(text.getText());
        }
    }
    class WindowAdapter1 extends WindowAdapter{
        public void windowClosing(WindowEvent e)
        {
            System.exit(0);
        }
    }
    public static void main(String args[])
    {
        ShowText1 show=new ShowText1();}
}
```

运行结果如图 11.5 所示。

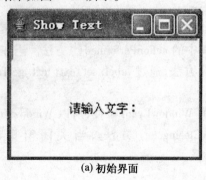
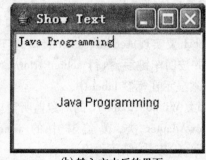

(a) 初始界面　　　　　　(b) 输入文本后的界面

图 11.5　运行结果

2. 编程实现：有一标题为"计算"的窗口，窗口的布局为 FlowLayout；有四个按钮，分别为"加"、"差"、"积"和"商"；另外，窗口中还有三个文本行，单击任一按钮，将两个文本行的数字进行相应的运算，在第三个文本行中显示结果。

解析：需定义"加"、"减"、"乘"、"除"4 个按钮，定义 3 个文本行分别显示参与运算的两个数据及其运算结果。为了直观起见,还定义了3个标签分别用于标示3个文本行中的信息。每个按钮的单击事件(ActionEvent)监听器都使用匿名类对象,匿名类需要实现 ActionListener 接口,实现其中的 actionPerformed( )方法。当用户单击某一按钮时,该按钮的事件监听器监听单击事件,便执行对应的 actionPerformed( )方法,对两个文本行中的数据进行相应的运算并将运算结果显示在第三个文本行中。

对窗口事件 WindowEvent 也使用匿名类对象作为监听器,该匿名类继承 WindowAdapter 类,覆盖其中的 windowClosing( )方法。当关闭窗口时,执行 windowClosing( )方法,结束程序。

```java
import java.awt.*;
import java.awt.event.*;
public class Calculator extends Frame {
    private Button plus;
    private Button minus;
    private Button multiply;
    private Button divide;
    private TextField num1;
    private TextField num2;
    private TextField result;
    public Calculator( )
    {
        super("计算");
        this.setLayout(new FlowLayout( ));
        num1 = new TextField(5);
        num2 = new TextField(5);
        result = new TextField(5);
        plus = new Button("加");
        minus = new Button("减");
        multiply = new Button("乘");
        divide = new Button("除");
        this.add(new Label("数字1:"));
        this.add(num1);
        this.add(new Label("数字2:"));
        this.add(num2);
        this.add(new Label("结果:"));
        this.add(result);
        this.add(result);
        this.add(plus);
        this.add(minus);
        this.add(multiply);
        this.add(divide);
        plus.addActionListener(new ActionListener( ) {
            public void actionPerformed(ActionEvent e) {
                double a = Double.parseDouble(num1.getText( ));
                double b = Double.parseDouble(num2.getText( ));
```

```java
        result.setText(Double.toString(a+b));
      }
    });
    minus.addActionListener(new ActionListener() {
      public void actionPerformed(ActionEvent e)
      {
        double a = Double.parseDouble(num1.getText());
        double b = Double.parseDouble(num2.getText());
        result.setText(Double.toString(a-b));
      }
    });
    multiply.addActionListener(new ActionListener() {
      public void actionPerformed(ActionEvent e)
      {
        double a = Double.parseDouble(num1.getText());
        double b = Double.parseDouble(num2.getText());
        result.setText(Double.toString(a*b));
      }
    });
    divide.addActionListener(new ActionListener() {
      public void actionPerformed(ActionEvent e)
      {
        double a = Double.parseDouble(num1.getText());
        double b = Double.parseDouble(num2.getText());
        result.setText(Double.toString(a/b));
      }
    });
    this.addWindowListener(new WindowAdapter() {
      public void windowClosing(WindowEvent e) {
        System.exit(0);
      }
    });
    this.setSize(150,200);
    this.setVisible(true);
  }
  public static void main(String[] args) {
    Calculator calcul = new Calculator();
  }
}
```

运行结果如图11.6所示。

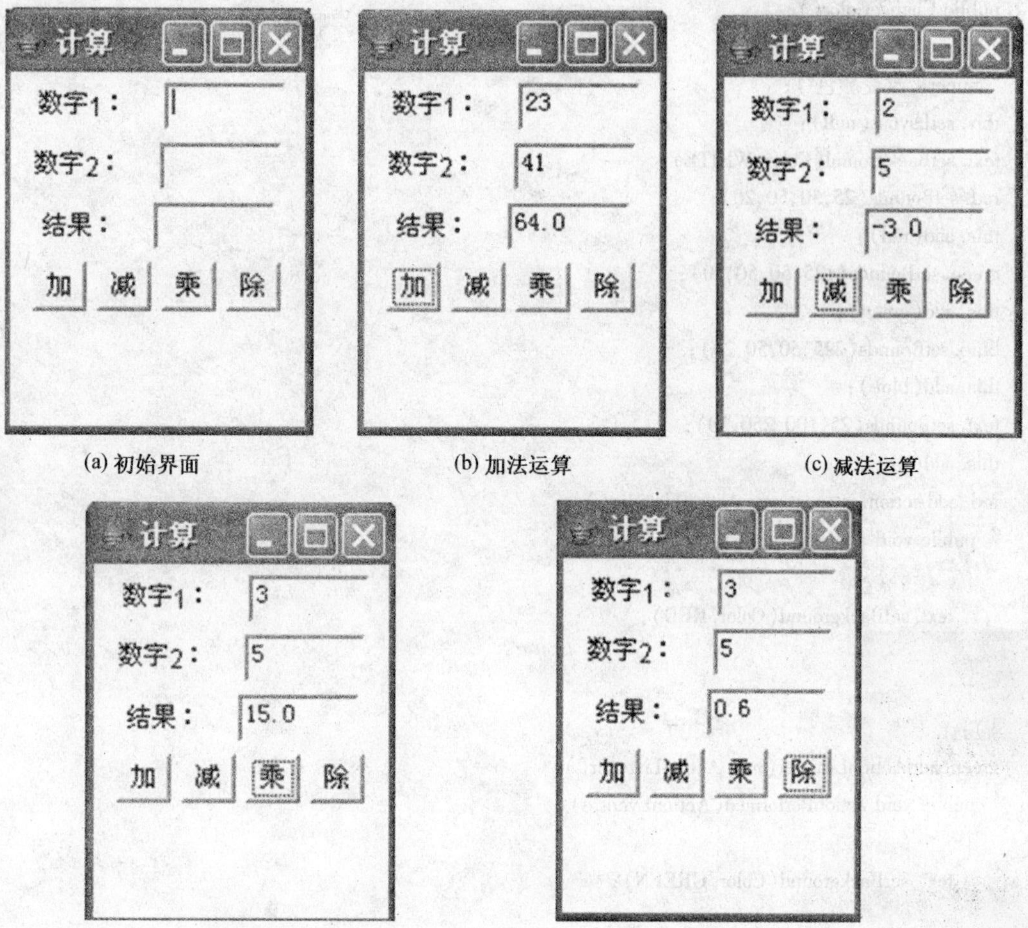

图 11.6 运行结果

(a) 初始界面　　(b) 加法运算　　(c) 减法运算　　(d) 乘法运算　　(e) 除法运算

3. 编写应用程序,有一个标题为"改变颜色"的窗口,窗口布局为 null,在窗口中有三个按钮和一个文本行,三个按钮的颜色分别是"红"、"绿"和"蓝",单击任一按钮,文本行的背景色更改为相应的颜色。

解析:需定义"红"、"绿"和"蓝"三个按钮,定义一个文本行。每个按钮的单击事件(ActionEvent)监听器都使用匿名类对象,匿名类需要实现 ActionListener 接口,实现其中的 actionPerformed()方法。当用户单击某一按钮时,该按钮的事件监听器监听到单击事件,便执行对应的 actionPerformed()方法,将文本行的背景色更改为相应的颜色。

对窗口事件 WindowEvent 也使用匿名类对象作为监听器,该匿名类继承 WindowAdapter 类,覆盖其中的 windowClosing()方法。当关闭窗口时,执行 windowClosing()方法,结束程序。

```
import java.awt.*;
import java.awt.event.*;
public class ChangeColor extends Frame{
    private Button red=new Button("红");
    private Button green=new Button("绿");
    private Button blue=new Button("蓝");
    private TextField text=new TextField();
```

```java
public ChangeColor()
{
    super("改变颜色");
    this.setLayout(null);
    text.setBackground(Color.WHITE);
    red.setBounds(25,50,50,20);
    this.add(red);
    green.setBounds(125,50,50,20);
    this.add(green);
    blue.setBounds(225,50,50,20);
    this.add(blue);
    text.setBounds(25,100,250,30);
    this.add(text);
    red.addActionListener(new ActionListener(){
        public void actionPerformed(ActionEvent e)
        {
            text.setBackground(Color.RED);

        }
    });
    green.addActionListener(new ActionListener(){
        public void actionPerformed(ActionEvent e)
        {
            text.setBackground(Color.GREEN);

        }
    });
    blue.addActionListener(new ActionListener(){
        public void actionPerformed(ActionEvent e)
        {
            text.setBackground(Color.BLUE);

        }
    });
    addWindowListener(new WindowAdapter(){
        public void windowClosing(WindowEvent e){
            System.exit(0);
        }
    });
    setSize(300,200);
    setVisible(true);
}
public static void main(String args[]){
    ChangeColor color=new ChangeColor();
```

}
}

运行结果如图 11.7 所示。

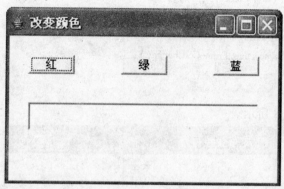

图 11.7 ChangeColor 运行界面

4. 编写一个简单的屏幕变色程序。当用户单击"变色"按钮时,窗口颜色就自动地变成一种颜色。

解析:程序中定义了按钮 change,其单击事件(ActionEvent)监听器使用实现 ActionListener 接口的 FrameColChange 类对象(this)。所以,FrameColChange 类实现了 actionPerformed( ) 方法。当用户单击 change 按钮时,事件监听器监听到单击事件,便执行 actionPerformed( ) 方法,通过"(int)(Math.random( )*1000)%256"产生 3 个 0~255 之间的随机整数,分别存放在 int 型变量 r,g 和 b 中,再通过"setBackground(new Color(r,g,b))"将 3 个随机整数对应的颜色设置为窗口的背景色。

```java
import java.awt.*;
import java.awt.event.*;
public class FrameColChange extends Frame implements ActionListener{
    private Button change=new Button("变色");
    public FrameColChange( )
    {
        super("屏幕变色程序");
        this.setLayout(new FlowLayout( ));
        this.add(change);
        change.addActionListener(this);
        addWindowListener(new WindowAdapter( ){
            public void windowClosing(WindowEvent e)
            {
                System.exit(0);
            }
        });
        setSize(300,200);
        setVisible(true);
    }
    public void actionPerformed(ActionEvent arg0){
        int r=(int)(Math.random( )*1000)%256;
```

```
            int g=(int)(Math.random()*1000)%256;
            int b=(int)(Math.random()*1000)%256;
            this.setBackground(new Color(r,g,b));
        }
        public static void main(String args[]){
            FrameColChange frame=new FrameColChange();
        }
    }
```

运行结果如图 11.8 所示。

图 11.8　FrameColChange 运行界面

5. 编写一个温度转换程序。用户在文本行中输入华氏温度,并按回车键,自动在两个文本中分别显示对应的摄氏和 K 氏温度。要求给文本行和标签添加相应的提示信息。

具体的计算公式为:

$$摄氏温度=5(华氏温度-32)/9$$

$$K 氏温度=摄氏温度+273$$

解析:需定义一个文本行用来输入华氏温度,再定义两个文本行分别用来显示摄氏温度和 K 氏温度。为直观起见,还定义了三个标签分别用来标示三个文本行中的信息。需要对输入华氏温度的文本行监听 ActionEvent 事件。事件监听器采用匿名类对象,匿名类需要实现 ActionListener 接口,实现其中的 actionPerformed( )方法。当用户按下回车键时,事件监听器监听到事件,便执行对应的 actionPerformed( )方法,计算相应的摄氏温度和 K 氏温度,并将它们分别显示在相应的文本行中。

运行结果如图 11.9 所示。

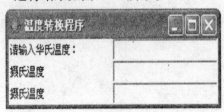

(a) 初始界面　　　　　　　　　　(b) 温度转换操作后的界面

图 11.9　运行界面

6. 创建一个有文本框和三个按钮的框架窗口程序,同时要求按下不同按钮时,文本框中能显示不同的文字。

```java
import java.awt.*;
import java.awt.event.*;
public class ActionEvent1
{
    private static Frame frame;            //定义为静态变量以便 main 使用
    private static Panel myPanel;          //该面板用来放置按钮组件
    private Button button1;                //这里定义按钮组件
    private Button button2;
    private Button button3;
    private TextField textfield1;          //定义文本框组件
    public ActionEvent1()                  //构造方法,建立图形界面
    {
        // 创建面板容器类组件
        myPanel = new Panel();
        // 创建按钮组件
        button1 = new Button("按钮 1");
        button2 = new Button("按钮 2");
        button3 = new Button("按钮 3");
        textfield1 = new TextField(30);
        MyListener myListener = new MyListener();
        // 建立一个 actionlistener 让两个按钮共享
        button1.addActionListener(myListener);
        button2.addActionListener(myListener);
        button3.addActionListener(myListener);
        textfield1.addActionListener(myListener);
        myPanel.add(button1);  // 添加组件到面板容器
        myPanel.add(button2);
        myPanel.add(button3);
        myPanel.add(textfield1);
    }
    //定义行为事件处理内部类,它实现 ActionListener 接口
    private class MyListener implements ActionListener
    {
        /*
        利用该内部类来监听所有行为事件源产生的事件
        */
        public void actionPerformed(ActionEvent e)
        {
            // 利用 getSource() 方法获得组件对象名
            // 也可以利用 getActionCommand() 方法来获得组件标识信息
            // 如 e.getActionCommand().equals("按钮 1")
            Object obj = e.getSource();
            if (obj == button1)
                textfield1.setText("按钮 1 被单击");
```

```java
        else if (obj==button2)
            textfield1.setText("按钮2被单击");
        else if (obj==button3)
            textfield1.setText("按钮3被单击");
        else
            textfield1.setText("");
    }
}
public static void main(String s[])
{
    ActionEvent1 ae=new ActionEvent1();  // 新建 ActionEvent1 组件
    frame=new Frame("ActionEvent1");  // 新建 Frame
    // 处理窗口关闭事件的通常方法(属于匿名内部类)
    frame.addWindowListener(new WindowAdapter(){
        public void windowClosing(WindowEvent e)
        {System.exit(0);}  });
    frame.add(myPanel);
    frame.pack();
    frame.setVisible(true);
}
}
```

11.3 上机实验

一、实验目的与意义

1. 熟悉 AWT 标签、文本框、文本行、按钮等组件的使用方法；
2. 熟悉事件处理方法。

二、实验内容

1. 设计一个简易计算器,如图 11.10 所示。在"操作数"标签右侧的两个文本行输入操作数,当单击"+,-,×,÷"按钮时,对两个操作数进行运算并将结果填入到"结果"标签右侧的文本行中。

2. 编写文本移动程序,窗口如图 11.11 所示。窗口中有 2 个文本区和 2 个按钮,文本区分别位于窗口的左边和右边区域,2 个按钮位于窗口的中间区域,当单击"→"按钮时,将左边文本区中选中的内容添加到右边文本区的末尾。当单击"←"按钮时,将右边文本区中选中的内容添加到左边文本区的末尾。

提示:文本区中,可以使用 getSelectedText() 方法获得通过鼠标拖动选中的文本,可以将"→"和"←"按钮放入 Panel 组件中,再将 Panel 组件加入窗口。

图 11.10　简易计算器界面

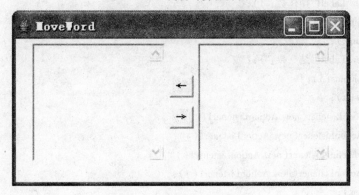

图 11.11　文本移动窗口

三、实验要求

1. 组件在窗口中的位置尽量按要求摆放；
2. 响应窗口的关闭操作事件。

11.4　程序代码

1. 简易计算器程序

```
import java.awt.*;
import java.awt.event.*;
public class Calculator1 extends Frame{
    private Button plus;
    private Button minus;
    private Button multiply;
    private Button divide;
    private TextField num1;
    private TextField num2;
    private TextField result;
public Calculator1( ){
    super("简易计算器");
```

```java
        this.setLayout(new FlowLayout());
        plus = new Button("+");
        minus = new Button("-");
        multiply = new Button(" * ");
        divide = new Button("/");
        num1 = new TextField();
        num2 = new TextField();
        result = new TextField();
        this.add(new Label("操作数:"));
        num1.setColumns(5);
        this.add(num1);
        this.add(new Label("操作数:"));
        num2.setColumns(5);
        this.add(num2);
        this.add(new Label("结  果:"));
        result.setColumns(5);
        this.add(result);
        plus.addActionListener(new ActionListener1());
        minus.addActionListener(new ActionListener1());
        multiply.addActionListener(new ActionListener1());
        divide.addActionListener(new ActionListener1());
        this.add(plus);
        this.add(minus);
        this.add(multiply);
        this.add(divide);
        this.addWindowListener(new WindowAdapter(){
            public void windowClosing(WindowEvent e){
                System.exit(0);
            }
        });
        this.setSize(150,200);
        this.setVisible(true);
    }
    class ActionListener1 implements ActionListener{
        public void actionPerformed(ActionEvent e){
            double a = Double.parseDouble(num1.getText());
            double b = Double.parseDouble(num2.getText());
            if(e.getSource() == plus)
                result.setText(Double.toString(a+b));
            if(e.getSource() == minus)
                result.setText(Double.toString(a-b));
            if(e.getSource() == multiply)
                result.setText(Double.toString(a * b));
            if(e.getSource() == divide)
```

```
            result.setText(Double.toString(a/b));
        }
    }
    public static void main(String args[]){
        Calculator1 calcul=new Calculator1();
    }
}
```

2. 文本移动程序

```
import java.awt.*;
import java.awt.event.*;
public class MoveWord extends Frame{
    private TextArea eastArea=new TextArea(7,20);
    private TextArea westArea=new TextArea(7,20);
    private Button toLeft=new Button("←");
    private Button toRight=new Button("→");
    public MoveWord(){
        super("MoveWord");
        this.setLayout(new FlowLayout());
        this.add(westArea);
        Panel pal=new Panel();
        pal.setLayout(new GridLayout(2,1,10,10));
        pal.add(toLeft);
        pal.add(toRight);
        toLeft.addActionListener(new Handler());
        toRight.addActionListener(new Handler());
        this.add(pal);
        this.add(eastArea);
        addWindowListener(new WindowAdapter(){
            public void windowClosing(WindowEvent e){
                System.exit(0);
            }
        });
        this.setSize(400,200);
        this.setVisible(true);
    }
}
class Handler implements ActionListener{
    public void actionPerformed(ActionEvent e)
    {
        String copyText="";
        if(e.getSource()==toLeft){
            copyText=eastArea.getSelectedText();
            westArea.append(copyText);
        }
        else
```

```
        }
            copyText = westArea.getSelectedText();
            eastArea.append(copyText);
        }
      }
    }
    public static void main(String[] args) {
      MoveWord word = new MoveWord();
    }
}
```

第12章

Swing 组件

12.1 典型例题解析

【例12.1】 编写应用程序实现:窗口有一个 JList 组件和一个 JLabel 组件,JList 组件显示若干大学名称,选定某个大学名称后,JLabel 组件显示选定大学的地址。

解析:定义一维字符串数组 name 和 address,分别用来存放4所大学的名称和对应的地址。定义 JList 对象 list,其中显示 name 中的4所大学名称。list 的 ListSelectionEvent 事件采用内部类 Handler 的对象作为监听器,Handler 类需要实现 ListSelectionListener 接口,实现其中的 valueChange( )方法。当用户选择 list 中的某个学校名称时,事件监听器监听到 ListSelectionEvent 事件,便执行 valueChanged( )方法,通过 list.getSelectedIndex( )获得 list 中被选中学校的序号,通过 address[list.getSelectedIndex( )]获得 list 中被选中学校对应的地址,通过 lbl.setText( )方法将对应的地址在 lbl 标签中显示。

要对窗口关闭操作进行响应,需要对其注册 WindowEvent 事件监听器。事件监听器由继承 WindowAdapter 适配器类的匿名类对象担任。匿名类覆盖了 WindowAdapter 中的 windowClosing( )方法,用户单击窗口关闭按钮时,结束程序运行。

```
import java.awt.*;
import java.awt.event.*;
import javax.swing.*;
import javax.swing.event.*;
public class ListExample extends JFrame{
private JList list;
private JLabel lbl;
private String[] name={"西安交通大学","西北大学","西北工业大学","第四军医大学"};
private String[] address={"兴庆路","大学南路","友谊西路","长乐东路"};
public ListExample(){
  super("ListExample");
  Container c=getContentPane();
  c.setLayout(new BorderLayout());
  lbl=new JLabel();
  c.add(lbl,BorderLayout.SOUTH);
  list=new JList(name);
```

```
list.setSelectionMode(ListSelectionModel.SINGLE_SELECTION);
c.add(list,BorderLayout.NORTH);
list.addListSelectionListener(new Handler());
addWindowListener(new WindowAdapter()｛
   public void windowClosing(WindowEvent e)｛
      System.exit(0);｝
｝);
setSize(250,170);
setVisible(true);
｝
public static void main(String args[ ])｛
   ListExample app=new ListExample();
｝
private class Handler implements ListSelectionListener｛
   public void valueChanged(ListSelectionEvent e)
   ｛String s;
    lbl.setText("地址:"+address[list.getSelectedIndex()]);｝
   ｝
｝
```

运行结果如图 12.1 所示。

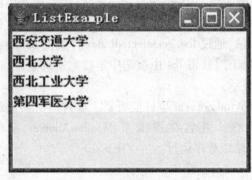
(a) 初始界面

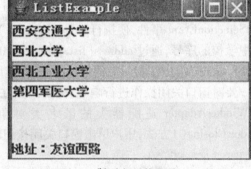
(b) 选择后的界面

图 12.1　ListExample 运行界面

【例 12.2】　编程实现:窗口中有两个组合框(JComBox 组件),分别显示字体名称和字体大小,还有一个标签,其标题为"字体及大小样例"。当用户从组合框中选择某字体名称和字体大小时,标签的标题就更改为相应的字体和大小。

解析:定义了一维字符串数组 names 和 sizes,分别用来存放字体名称和字体大小。定义了两个 JComboBox 组件 jName 和 jSize,分别显示 names 和 sizes 中的字体名称和字体大小。jName 和 jSize 的 ItemEvent 事件(选择某一选项时触发)采用内部类 Handler 的对象作为监听器,Handler 类需要实现 ItemListener 接口,实现其中的 itemStateChanged() 方法。当用户选择 jName 或 jSize 中的某个选项时,事件监听器监听到 itemEvent 事件,便执行 itemStateChanged() 方法。

在 itemStateChanged() 中,通过 getSource() 获得触发 ItemEvent 事件的事件源。如果事件源是 jName(用户选择了某个字体名称),通过 jName.setSelectedIndex() 获得用户选择的字体名称对应的序号,通过 names[jName.getSelectedIndex()] 获得用户所选择的字体名称,并将所选择的字体名称存放在字符串变量 fontName 中;同理,如果事件源是 jSize(用户选择了某个字体大小),通过 jSize.getSelectedIndex() 获得用户选择的字体大小对应的序号,通过 size[jSize.getSelectedIndex()] 获得用户所选择的字体大小(字符串型),通过 Inter.parseInt(jSize[jSize.getSelectedIndex()]) 将用户所选择的字体大小由字符串型转换成整型。通过语句 font = new Font(fontName,Font.PLAIN,fontSize) 创建用户所选择的字体对象(所选择的字体名称和大小),并存放在 font 中。最后,通过 lbl.setFont(font) 将标签 lbl 的标题设置为用户所选择的字体。

运行结果如图 12.2 所示。

(a) 初始界面　　　　　　　　　　(b) 选择了字体后的界面

图 12.2　FontExample 运行界面

【例 12.3】　编写程序实现:通过文本行输入学生姓名,通过单选按钮选择性别,通过复选框选择课程,并在文本框中显示所填写及选择的信息。请自行安排版面,使其美观。

解析:文本行 txf 供输入学生姓名,man、woman 和 radioGroup 组成单选按钮供选择性别,math 和 chinese 组成的复选框供选择课程,文本框 txa 用来显示所填写及选择的信息。

当单击命令按钮 ok 时,文本框 txa 显示目前所填写及选择的最新信息;当单击命令按钮 cancel 时,结束程序运行。

ok 和 cancel 的 ActionEvent 事件(单击某一按钮时触发)采用内部类 Handler 的对象作为监听器,Handler 类需要实现 ActionListener 接口,实现其中的 actionPerformed() 方法。当用户单击某一按钮时,事件监听器监听到 ActionEvent 事件,便执行 actionPerformed() 方法。

在 actionperformed() 中,通过 getSource() 获得触发 ActionEvent 事件的事件源。如果事件源是 ok,将用户填写或选择的信息分别存放在字符串数组 str 的元素 str[0]、str[1]、str[2] 和 str[3] 中。通过 txf.getText() 取得 txf 中所输入的姓名,将选择结果存放于 str[1] 中;如果 main.isSelected() 的返回值是 true,表明选择了"男",否则选择了"女",将选择结果存放于 str[1];如果 math.isSelected() 的返回值是 false,表明选择了"数学",将其存放于 str[2] 中;如果 chinese.isSelected() 的返回值是 true,表明选择了"语文",将其存放于 str[3] 中;通过 for 循环语句,将存放于 str[0]、str[1]、str[2] 和 str[3] 中的信息分行存放于字符串变量 output 中;最后通过 txa.setText(output) 在文本区 txa 中显示所填写及选择的信息。如果事件源是 cancel,结束程序运行。

```
import java.awt.*;
import java.awt.event.*;
```

```java
import javax.swing.*;
public class MultiComponent extends JFrame{
  private JLabel name=new JLabel("姓名:");
  private JLabel sex=new JLabel("性别:");
  private JTextField txf=new JTextField(15);
  private JRadioButton man=new JRadioButton("男",true);
  private JRadioButton woman=new JRadioButton("女",false);
  private ButtonGroup radioGroup;
  private JCheckBox math=new JCheckBox("数学");
  private JCheckBox chinese=new JCheckBox("语文");
  private JButton ok=new JButton("确定");
  private JButton cancel=new JButton("取消");
  private String[] str=new String[4];
  private String output="";
  private JTextArea txa=new JTextArea(5,20);
  public MultiComponent(){
    super("MultiComponent");
    Container c=getContentPane();
    c.setLayout(new FlowLayout());
    c.add(name);
    c.add(txf);
    c.add(sex);
    c.add(man);
    c.add(woman);
    radioGroup=new ButtonGroup();
    radioGroup.add(man);
    radioGroup.add(woman);
    ok.addActionListener(new handlel());
    cancel.addActionListener(new handlel());
    c.add(math);
    c.add(chinese);
    c.add(ok);
    c.add(cancel);
    c.add(txa);
    setSize(230,200);
    setVisible(true);
  }
  public static void main(String args[]){
    MultiComponent app=new MultiComponent();
    app.addWindowListener(new WindowAdapter(){
      public void windowClosing(WindowEvent e){
        System.exit(0);
      }
    });
```

```
    }
    private class handlel implements ActionListener{
        public void actionPerformed( ActionEvent e){
            if( e. getSource( )= = ok)
            {
                str[0]="姓名:"+txf. getText( );
                if( man. isSelected( ))
                    str[1]="性别:男";
                else
                    str[1]="性别:女";
                if( math. isSelected( ))
                    str[2]="mathematics";
                else
                    str[2]="";
                if( chinese. isSelected( ))
                    str[3]="chinese";
                else
                    str[3]="";
                output="";
                for( int i=0;i<2;i++)
                output=output+str[i]+"\n";
                output=output+"所选课程如下:"+"\n";
                for( int i=2;i<4;i++)
                    output=output+str[i]+"\n";
                txa. setText( output);

            }
            if( e. getSource( )= = cancel)
                System. exit(0);
        }
    }
}
```

运行结果如图 12.3 所示。

(a)初始界面　　　　　　(b)输入及选择后的界面

图 12.3　MultiCompontent 运行界面

12.2 课后习题解答

1. 简述 AWT 组件和 Swing 组件的异同。

答：Swing 包含了大部分与 AWT 对应的组件，Swing 组件的用法与 AWT 组件基本相同，大多数 AWT 组件只要在其类名前加 J 即可转换成 Swing 组件。

Swing 组件与 AWT 最大的不同是：Swing 组件在实现时不包含任何本地代码，因此 Swing 组件可以不受硬件平台的限制，而具有更多的功能。不包含本地代码的 Swing 组件称为"轻量级"组件，而包含本地代码的 AWT 组件称为"重量级"组件。在 Java 2 平台上推荐使用 Swing 组件。

Swing 组件比 AWT 组件拥有更多的功能。例如，Swing 中的按钮和标签不仅可以显示文本信息，还可以显示图标，或同时显示文本和图标；大多数 Swing 组件都可以添加边框；Swing 组件可以具有任意开关，而不仅局限于长方形。

2. 编制程序实现：在 JTextField 中输入文本，单击按钮后，将所输文本添加到 JTextArea 中。

答：程序中定义了一个文本行、一个文本区和一个按钮。调用 JFrame 类的 getContentPane() 方法获得窗口的内容窗格，将其赋予 Container 类对象 pane，通过 pane 向窗口中添加组件，采用 Borderlayout 布局。对按钮的 ActionEvent 事件采用匿名类对象作为监听器，匿名类需要实现 ActionListener 接口，实现其中 actionPerformed() 方法。当用户单击按钮时，事件监听器监听到单击事件，便执行 actionPerformed() 方法，通过 getText() 将文本行的文本存放于字符串变量 text 中，再通过 append() 方法将 text 中的文本添加到文本区的末尾。

addtext(AddText 的对象)的窗口事件 WindowEvent 采用匿名类对象作为监听器，匿名类需要继承 WindowAdapter 类，覆盖其中的 windowClosing() 方法。

```java
import javax.swing.*;
import java.awt.*;
import java.awt.event.*;
public class AddText extends JFrame{
    private JTextField textField;
    private JTextArea textArea;
    private JButton button;
    public AddText()
    {
        super("Add text to textarea");
        Container pane = getContentPane();
        pane.setLayout(new BorderLayout(5,5));
        textField = new JTextField(20);
        textArea = new JTextArea(10,20);
        button = new JButton("单击按钮添加");
        button.addActionListener(new ActionListener(){
            public void actionPerformed(ActionEvent e){
                String text = textField.getText()+" ";
                textArea.append(text);
            }
        }
```

```
        });
        pane.add(textField,BorderLayout.NORTH);
        pane.add(button,BorderLayout.CENTER);
        pane.add(textArea,BorderLayout.SOUTH);
        setSize(300,300);
        this.show();
    }
    public static void main(String[] args)
    {
        AddText addText=new AddText();
        addText.addWindowListener(new WindowAdapter(){
            public void WindowClosing(WindowEvent e)
            {
                System.exit(0);
            }
        });
    }
}
```

运行结果如图 12.4 所示。

(a) 初始界面　　　　　　　　(b) 输入及添加操作后的界面

图 12.4　运行界面

3. 编写应用程序实现：窗口取默认布局——BorderLayout 布局，北面添加 JComboBox 组件，该组件有 6 个选项，分别表示 6 种商品名称。在中心添加一个文本区，当选择 JComboBox 组件中的某个选项后，文本区显示该商品的价格和产地信息。

答：定义了一维字符串数组 names，用来存放 6 种书名和 JcomboBox 组件中的标题"请选择要查询的商品名称"。定义了二维字符串数组 infos，用来存放 names 中 6 种书的信息和一行空信息。每种书的信息用二维数组中的一行表示，分别表示书名、出版社和价格，空信息对应 JComboBox 组件中的标题"请选择要查询的商品名称"。

定义了 JComboBox 对象 list，其中显示 names 中的 6 种书名和标题"请选择要查询的商品名称"。List 的 ItemEvent 事件采用匿名类对象作为监听器，匿名类需要实现 ItemListener 接口，实现其中的 itemStateChanged() 方法。当用户选择 list 中的某个书名时，事件监听器监听到 ItemEvent 事件,便执行 itemStateChanged( )方法,通过 list. getSelectedIndex( )获得 list 中被选中书名的序号,并存放在变量 index 中。通过 infos[index][0]、infos[index][1]和 infos[index][2]分别获得 list 中被选中书在 infos 中对应的书名、出版社和价格,通过 info. setText( )和 info. append( )方法分别将对应的书名、出版社名和价格在 JtextArea 对象 info 中显示。

```java
import java.awt. * ;
import javax.swing. * ;
import java.awt.event. * ;
public class Information extends JFrame{
    private JComboBox list;
    private JTextArea info;
    private String names[ ] = {"请选择要查询的商品名称","Linux 程序设计",
       "Windows 核心编程","操作系统概念","UNIX 技术手册","计算机操作系统",
       "Linux 系统开发员"};
    private String infos[ ][ ] = {
       {"","",""},
       {"Linux 程序设计","人民邮电出版社","89.00"},
       {"Windows 核心编程","机械工业出版社","86.00"},
       {"操作系统概念","高等教育出版社","55.00",
        "UNIX 技术手册","中国电力出版社","69.00"},
        {"计算机操作","清华大学出版社","21.00"},
        {"Linux 系统开发员","机械工业出版社","23.00"}};
    public Information( )
    {
        super("Information of merchandise");
        Container pane = this.getContentPane( );
        pane.setLayout(new BorderLayout( ));
        list = new JComboBox(names);
        info = new JTextArea(5,20);
        pane.add(list,BorderLayout.NORTH );
        pane.add(info,BorderLayout.CENTER );
        list.addItemListener(new ItemListener( ){
            public void itemStateChanged(ItemEvent e )
            {
                int index = list.getSelectedIndex( );
                info.setText("商品名:"+infos[index][0]+"\n");
                info.append("出版社"+infos[index][0]+"\n");
                info.append("市场价:"+infos[index][2]+"\n");
            }
        });
        this.setSize(250,300);
        this.setVisible(true);
    }
```

```
        public static void main(String args[ ])
        {
            Information information = new Information( );
            information.addWindowListener( new WindowAdapter( ){
                public void winowClosing( WindowEvent e)
                {
                    System.exit(0);
                }
            });
        }
```

运行结果如图 12.5 所示。

(a) 初始界面

(b) 选择操作后的界面

图 12.5 运行界面

4. 编写"猜数游戏"程序。系统自动生成一个 1~200 之间的随机整数,并在屏幕显示:"有一个数,在 1~200 之间。猜猜看,这个数是多少?"

答:用户在 JTextField 输入一个数,并按回车键。如果输入的数过大,JLabel 背景变红,同时显示"太大";如果输入的数过小,JLabel 背景变蓝,同时显示"太小";如果输入的数正好,JLabel 背景变白,同时显示"恭喜你!答对了!"。

解析:利用文本行 input 来输入数据,利用标签 message 来显示"太大"、"太小"等信息,机器产生的随机数放在变量 guessNum 中。Input 的 ActionEvent 事件采用匿名类对象作为监听器,匿名类需要实现 ActionListener 接口,实现其中的 actionPerformed( )方法。当用户在 input 中输入数据并按回车键时,事件监听器监听到 ActionEvent 事件,便执行 actionPerformed( )方法,将输入的数据放入变量 guessed 中,并以 guessed 作为参数调用 guess( )方法,将 guessed 与 guessNum 进行对比。如果 guessed < guessNum,guess( )返回 -1;如果 guessed > guessNum,guess( )返回 1;否则,返回 0。根据 guess( )的返回值,在 message 中显示"太小"或"太大"信息,直到用户输入的数据和产生的随机数相等,在 message 中显示"恭喜你!答对了!"。

```
import java.awt.*;
import javax.swing.*;
import java.awt.event.*;
public class Game extends JFrame {
```

```java
        private int guessNum;
        private JLabel message;
        private JTextField input;
        public Game() {
            super("Guess Game");
            Container pane=this.getContentPane();
            pane.setLayout(new BorderLayout());
            guessNum=((int)(Math.random()*200))+1;
            input=new JTextField();
            message=new JLabel("有1个数,在1~200之间。猜猜看,这个数是多少?");
            message.setHorizontalAlignment(SwingConstants.CENTER);
            message.setBackground(Color.WHITE);
            message.setFont(new Font("TimesRoman", Font.PLAIN, 20));
            pane.add(input, BorderLayout.NORTH);
            pane.add(message, BorderLayout.CENTER);
            input.addActionListener(new ActionListener() {
                public void actionPerformed(ActionEvent e) {
                    int guessed=Integer.parseInt(input.getText());
                    if (guess(guessed)==-1) {
                        message.setBackground(Color.BLUE);
                        message.setText("太小");
                        input.setText("");
                    } else if (guess(guessed)==1) {
                        message.setBackground(Color.RED);
                        message.setText("太大");
                        input.setText("");
                    } else if (guess(guessed)==0) {
                        message.setBackground(Color.WHITE);
                        message.setText("恭喜你!答对了!");
                    }
                }
            });
            this.setSize(500, 300);
            this.setVisible(true);
        }
        private int guess(int num) {
            if (num < this.guessNum)
                return -1;
            else if (num > this.guessNum)
                return 1;
            else
                return 0;
        }
        public static void main(String args[]) {
```

```
    Game game=new Game();
    game.addWindowListener(new WindowAdapter() {
      public void WindowClosing(WindowEvent e) {
        System.exit(0);
      }
    });
  }
}
```

5. 编写一个简单的个人简历程序。可以通过文本行输入姓名,通过单选按钮选择性别,通过组合框选择籍贯,通过列表框选择文化程度,请自行安排版面,使其美观。

答:文本行 nameInput 供输入姓名。Male,female 和 sexSelect 组成单选按钮供选择性别。字符串数组 privince 中存放多个省市名,组合框 provinceBox 显示 province 中存放的省市名,供选择籍贯用。字符串数组 degree 中存放各种学历名,列表框 degreeList 显示 degree 中存放的学历名,供选择学历用。文本框 txa 用来显示所填写及选择的信息。

当单击命令按钮 ok 时,文本框 txa 显示目前所填写及选择的最新信息;当单击命令按钮 cancel 时,结束程序运行。

ok 和 cancel 的 ActionEvent 事件(单击某一按钮时触发)采用内部类 Handle1 的对象作为监听器,Handle1 类需要实现 ActionListener 接口,实现其中的 actionPerformed()方法。当用户单击某一按钮时,事件监听器监听到 ActioonEvent 事件,便执行 actionPerformed()方法。

在 actionPerformed()中,通过 getSource()获得触发 ActionEvent 事件的事件源。如果事件源是 ok,将用户填写或选择的信息分别存放在字符串数组 str 的元素 str[0],str[1],str[2] 和 str[3]中。通过 nameInput.getText()取得 nameInput 中所输入的姓名,将其存放于 str[1]中;如果 male.isSelected()的返回值是 true,表明选择了"男",否则选择了"女",将选择结果存放于 str[1]中;通过 provinceBox.getSelectedIndex()获得用户在 provinceBox 中选择的籍贯中的序号,通过 province[provinceBox.getSelectedIndex()]获得用户所选择的籍贯,并将其存放于 str[2]中;通过 province[provinceBox.getSelectedIndex()]获得用户所选择的文化程度,并将其放于 str[3]中;通过 for 循环语句,将存放于 str[0],str[1],str[2] 和 str[3]中的信息分行存放于字符串变量 output 中;最后,通过 txa.setText(output)在文本框 txa 中显示所填写及选择的信息。如果事件源是 cancel,结束程序运行。

为了使窗口内组件排列整齐、美观,程序中使用了多个 JPanel(面板)组件,包括 namePan,sexPan,provincePan,degreePan 和 buttonPan。类似于 JFrame,JPanel 也是一种容器,可以包含 JButton,JTextField 和 JList 等组件,可以设置版面。将其他组件放入 JPanel 组件后,再将 JPanel 组件嵌入窗口,易于实现窗口内组件的合理布局。

```
import java.awt.*;
import javax.swing.*;
import java.awt.event.*;
public class Resume extends JFrame{
  private JLabel name;
  private JTextField nameInput;
  private JLabel sex;
  private JRadioButton male;
```

```java
    private JRadioButton female;
    private ButtonGroup sexSelect;
    private JLabel provinceLab;
    private String province[] = {"北京市","陕西省","河南省"};
    private JComboBox provinceBox;
    private JLabel degreeLab;
    private String degree[] = {"中学","本科","硕士","博士","其他"};
    private JList degreeList;
    private JButton ok,cancel;
    private String[] str = new String[4];
    private String output = "";
    private JTextArea txa;
    public Resume(){
        super("简单的个人简历程序");
        Container c = this.getContentPane();
        c.setLayout(new FlowLayout());
        name = new JLabel("姓名:");
        name.setHorizontalAlignment(SwingConstants.CENTER);
        nameInput = new JTextField(8);
        sex = new JLabel("性别:");
        sex.setHorizontalAlignment(SwingConstants.CENTER);
        male = new JRadioButton("男",true);
        female = new JRadioButton("女",false);
        sexSelect = new ButtonGroup();
        sexSelect.add(male);
        sexSelect.add(female);
        provinceLab = new JLabel("籍贯:");
        provinceLab.setHorizontalAlignment(SwingConstants.CENTER);
        provinceBox = new JComboBox(province);
        degreeLab = new JLabel("文化程度:");
        degreeLab.setHorizontalAlignment(SwingConstants.CENTER);
        degreeList = new JList(degree);
        degreeList.setVisibleRowCount(2);
        ok = new JButton("确定");
        cancel = new JButton("取消");
        txa = new JTextArea(5,20);
        JPanel namePan = new JPanel();
        namePan.add(name);
        namePan.add(nameInput);
        c.add(namePan);
        JPanel sexPan = new JPanel();
        sexPan.add(sex);
        sexPan.add(male);
        sexPan.add(female);
```

```java
        c.add(sexPan);
        JPanel provincePan=new JPanel();
        provincePan.add(provinceLab);
        provincePan.add(this.provinceBox);
        c.add(provincePan);
        JPanel degreePan=new JPanel();
        degreePan.add(degreeLab);
        degreePan.add(degreeList);
        c.add(degreePan);
        JPanel buttonPan=new JPanel();
        buttonPan.add(ok);
        buttonPan.add(cancel);
        ok.addActionListener(new Handle1());
        cancel.addActionListener(new Handle1());
        c.add(buttonPan);
        c.add(txa);
        this.setSize(350,350);
        this.setVisible(true);
    }
    public static void main(String args[]){
        Resume resume=new Resume();
        resume.addWindowListener(new WindowAdapter(){
            public void windowClosing(WindowEvent e)
            {
                System.exit(0);
            }
        });
    }
    private class Handle1 implements ActionListener{
        public void actionPerformed(ActionEvent e)
        {
            if(e.getSource()==ok)
            {
                str[0]="姓名:"+nameInput.getText();
                if(male.isSelected())
                    str[1]="性别:男";
                else
                    str[1]="性别:女";
                str[2]="籍贯:"+province[provinceBox.getSelectedIndex()];
                str[3]="文化程度:"+degree[degreeList.getSelectedIndex()];
                output="";
                for(int i=0;i<4;i++)
                    output=output+str[i]+"\n";
                txa.setText(output);
```

```
        }
        if( e. getSource( )= =cancel)
          System. exit(0);
      }
    }
}
```

运行结果如图 12.6 所示。

(a)初始界面　　　　　　　　　　(b)输入及选择操作后的界面

图 12.6　运行界面

6. 创建一个带有多级菜单系统的框架窗口程序,要求每单击一个菜单项,就弹出一个相对应的信息提示框。

```
import java. awt. *;
import java. awt. event. *;
import javax. swing. *;
class DialogFrame extends JFrame
    implements ActionListener
{   public DialogFrame( )
    {   setTitle("练习 11");
        setSize(300,300);
        addWindowListener( new WindowAdapter( )
        {   public void windowClosing( WindowEvent e)
            {   System. exit(0);
            }
        });
        JMenuBar mbar=new JMenuBar( );
        setJMenuBar(mbar);
        JMenu fileMenu=new JMenu("文件");
        JMenu s=new JMenu("特殊功能");
        s. add(specials[0]);
        specials[0]. addActionListener(this);
        s. add(specials[1]);
        specials[1]. addActionListener(this);
```

```java
        mbar.add(fileMenu);
        fileMenu.add(s);
        aboutItem = new JMenuItem("关于");
        aboutItem.addActionListener(this);
        fileMenu.add(aboutItem);
        exitItem = new JMenuItem("退出");
        exitItem.addActionListener(this);
        fileMenu.add(exitItem);
    }
    public void actionPerformed(ActionEvent evt)
    {   Object source = evt.getSource();
        if(source == aboutItem)
        {   if (dialog1 == null) // first time
                dialog1 = new AboutDialog(this,"关于");
            dialog1.show();
        }
        else if(source == exitItem)
        {   System.exit(0);
        }
        else if(source == specials[0])
        {   if (dialog2 == null) // first time
                dialog2 = new AboutDialog(this,"功能1");
            dialog2.show();
        }
        else if(source == specials[1])
        {   if (dialog3 == null) // first time
                dialog3 = new AboutDialog(this,"功能2");
            dialog3.show();
        }
    }
    private AboutDialog dialog1;
    private AboutDialog dialog2;
    private AboutDialog dialog3;
    private JMenuItem aboutItem;
    private JMenuItem exitItem;
    private JMenuItem[] specials = {
        new JMenuItem("功能1"),
        new JMenuItem("功能2")
    };
}
class AboutDialog extends JDialog
{   public AboutDialog(JFrame parent, String title)
    {   super(parent, title, true);
        JPanel p2 = new JPanel();
```

```java
            JButton ok = new JButton("Ok");
            p2.add(ok);
            getContentPane().add(p2, "Center");
            ok.addActionListener(new ActionListener()
                { public void actionPerformed(ActionEvent evt)
                    { setVisible(false); }
                });
            setSize(250, 150);
        }
    }
    public class DialogTest {
        public static void main(String[] args)
        { JFrame f = new DialogFrame();
            f.show();
        }
    }
}
```

7. 请分别用 AWT 及 Swing 组件来设计实现计算器程序,要求能完成简单四则运算。
答:下面给出 Swing 版的参考程序,AWT 版的类似。

```java
import java.awt.*;
import java.awt.event.*;
import javax.swing.*;
class CalculatorPanel extends Jpanel implements ActionListener
{   public CalculatorPanel()
    {   setLayout(new BorderLayout());
        display = new JTextField("0");
        display.setEditable(false);
        add(display, "North");
        JPanel p = new JPanel();
        p.setLayout(new GridLayout(4, 4));
        String buttons = "789/456*123-0.=+";
        for (int i=0; i < buttons.length(); i++)
            addButton(p, buttons.substring(i, i+1));
        add(p, "Center");
    }
    private void addButton(Container c, String s)
    {   JButton b = new JButton(s);
        c.add(b);
        b.addActionListener(this);
    }
    public void actionPerformed(ActionEvent evt)
    {   String s = evt.getActionCommand();
        if ('0' <= s.charAt(0) && s.charAt(0) <= '9'
            || s.equals("."))
        {   if (start) display.setText(s);
```

```java
            else display.setText(display.getText()+s);
            start=false;
         }
         else
         {  if (start)
            {  if (s.equals("-"))
                  { display.setText(s); start=false; }
               else op=s;
            }
            else
            {  double x =
                  Double.parseDouble(display.getText());
               calculate(x);
               op=s;
               start=true;
            }
         }
      }
      public void calculate(double n)
      {  if (op.equals("+")) arg+=n;
         else if (op.equals("-")) arg -=n;
         else if (op.equals("*")) arg *=n;
         else if (op.equals("/")) arg /=n;
         else if (op.equals("=")) arg=n;
         display.setText(""+arg);
      }
      private JTextField display;
      private double arg=0;
      private String op="=";
      private boolean start=true;
}
class CalculatorFrame extends JFrame
{  public CalculatorFrame()
   {  setTitle("Calculator");
      setSize(200, 200);
      addWindowListener( new WindowAdapter()
         {  public void windowClosing(WindowEvent e)
            {  System.exit(0);
            }
         });
      Container contentPane=getContentPane();
      contentPane.add(new CalculatorPanel());
   }
}
```

```java
public class Calculator
{   public static void main(String[ ] args)
    {   JFrame frame=new CalculatorFrame();
        frame.show();
    }
}
```

12.3 上机实验

一、实验目的与意义

1. 熟悉 Swing 组件的用法；
2. 熟悉事件处理方法。

二、实验内容

1. 编写"背单词"程序。系统从词库中随机抽取英文单词，通过一个 JLabel 组件显示对应的中文，让用户在 JTextField 组件中输入英文单词。如果用户输入的英文单词出错，按回车键后，系统在另一个 JLabel 组件显示"对不起！答错了！"，直到用户输入正确的英文单词，按回车键后，系统显示"恭喜你！答对了！"。该过程可以持续进行，直到用户终止程序。

2. 编写一个简单的个人简历录入程序。可以通过文本行输入姓名，通过单选按钮选择性别，通过组合框选择籍贯和文化程度，并在文本框中显示所填写及选择的信息。请自行安排版面，使其美观。

三、实验要求

1. 合理布局组件在窗口中的位置，使界面美观；
2. 响应窗口的关闭程序事件。

12.4 程序代码

1. 背单词程序

```java
import java.awt.*;
import javax.swing.*;
import java.awt.event.*;
public class WordStudy extends JFrame{
    private int index;
    private String word,guessedWord,character;
    private String[ ] words={"china","study","program","desk","room","computer"};
    private String[ ] characters={"中国","学习","程序","桌子","房间","计算机"};
    private JLabel lbl1,lbl2;
    private JTextField input;
    public WordStudy(){
```

```java
    super("Guess Game");
    Container c = this.getContentPane();
    c.setLayout(new FlowLayout());
    lbl1 = new JLabel();
    lbl2 = new JLabel("输入英文单词");
    input = new JTextField(15);
    c.add(lbl1);
    c.add(lbl2);
    c.add(input);
    input.addActionListener(new Handler());
    create();
    this.setSize(250,150);
    this.setVisible(true);
}
class Handler implements ActionListener{
    public void actionPerformed(ActionEvent e){
        guessedWord = (input.getText()).toLowerCase();
        if(guessedWord.equals(word)){
            lbl2.setBackground(Color.BLUE);
            lbl2.setText("恭喜你!答对了!");
            input.setText("");
            create();
        }
        else
        {
            lbl2.setBackground(Color.WHITE);
            lbl2.setText("对不起,答错了!");
            input.setText("");
        }
    }
}
public void create(){
    index = ((int)(Math.random() * 10));
    character = characters[index];
    word = words[index].toLowerCase();
    lbl1.setText(character);
}
public static void main(String[] args) {
    WordStudy app = new WordStudy();
    app.addWindowListener(new WindowAdapter(){
        public void WindowClosing(WindowEvent e)
        {
            System.exit(0);
```

        }
    });
    }
}

2. 简历录入程序

```java
import java.awt.*;
import javax.swing.*;
import java.awt.event.*;
public class SimpleResume extends JFrame{
    private JLabel name;
    private JTextField nameInput;
    private JLabel sex;
    private JRadioButton male;
    private JRadioButton female;
    private ButtonGroup sexSelect;
    private JLabel provinceLab;
    private String province[] = {"北京市","上海市","重庆市"};
    private JComboBox provinceBox;
    private JLabel degreeLab;
    private String degree[] = {"本科","硕士","博士","其他"};
    private JComboBox degreeList;
    private JButton ok,cancel;
    private String[] str = new String[4];
    private String output = "";
    private JTextArea txa;
    public SimpleResume(){
        super("简单的个人简历程序");
        Container c = this.getContentPane();
        c.setLayout(new FlowLayout());
        name = new JLabel("姓名:");
        name.setHorizontalAlignment(SwingConstants.CENTER);
        nameInput = new JTextField(8);
        sex = new JLabel("性别:");
        sex.setHorizontalAlignment(SwingConstants.CENTER);
        male = new JRadioButton("男",true);
        female = new JRadioButton("女",false);
        sexSelect = new ButtonGroup();
        sexSelect.add(male);
        sexSelect.add(female);
        provinceLab = new JLabel("籍贯:");
        provinceLab.setHorizontalAlignment(SwingConstants.CENTER);
        provinceBox = new JComboBox(province);
        degreeLab = new JLabel("文化程度:");
        degreeLab.setHorizontalAlignment(SwingConstants.CENTER);
```

```java
        degreeList = new JComboBox(degree);
        ok = new JButton("确定");
        cancel = new JButton("取消");
        txa = new JTextArea(5,20);
        JPanel namePan = new JPanel();
        namePan.add(name);
        namePan.add(nameInput);
        c.add(namePan);
        JPanel sexPan = new JPanel();
        sexPan.add(sex);
        sexPan.add(male);
        sexPan.add(female);
        c.add(sexPan);
        JPanel provincePan = new JPanel();
        provincePan.add(provinceLab);
        provincePan.add(this.provinceBox);
        c.add(provincePan);
        JPanel degreePan = new JPanel();
        degreePan.add(degreeLab);
        degreePan.add(degreeList);
        c.add(degreePan);
        JPanel buttonPan = new JPanel();
        buttonPan.add(ok);
        buttonPan.add(cancel);
        ok.addActionListener(new Handle1());
        cancel.addActionListener(new Handle1());
        c.add(buttonPan);
        c.add(txa);
        this.setSize(350,280);
        this.setVisible(true);
    }
    public static void main(String[] args) {
        SimpleResume resume = new SimpleResume();
        resume.addWindowListener(new WindowAdapter() {
            public void WindowClosing(WindowEvent e)
            {
                System.exit(0);
            }
        });
    }
    private class Handle1 implements ActionListener {
        public void actionPerformed(ActionEvent e)
        {
            if(e.getSource() == ok) {
```

```
        str[0]="姓名:"+nameInput.getText();
        if(male.isSelected())
            str[1]="性别:男";
        else
            str[1]="性别:女";
        str[2]="籍贯:"+province[provinceBox.getSelectedIndex()];
        str[3]="文化程度:"+degree[degreeList.getSelectedIndex()];
        output="";
        for(int i=0;i<4;i++)
            output=output+str[i]+"\n";
        txa.setText(output);
    }
    if(e.getSource()==cancel)
        System.exit(0);
    }
  }
}
```

第13章

Applet 程序

13.1 典型例题解析

【例13.1】 实现 Applet 程序中的单击按钮事件响应。编写 Applet 程序实现四则运算,单击+,-,*,/按钮,对两个输入数据执行对应的运算,并显示运算结果。

解析:单击按钮产生 ActionEvent 事件,Applet 响应这种事件与 JFrame 中响应这种事件的方法是一样的,也包括实现 ActionListener 接口定义监听类,注册监听对象等操作。

文本框对象 jt1,jt2,jt3 和按钮对象 b1,b2,b3,b4 在 init() 中创建,actionPerformed() 引用了这些对象,因此必须把这些对象声明成类的成员变量。

程序源代码:

```java
// Example_Applet13_1.java
import java.awt.*;
import java.awt.event.*;
import javax.swing.*;

public class Example_Applet13_1 extends JApplet implements ActionListener
{
    JTextField jt1,jt2,jt3;
    JButton b1,b2,b3,b4;
    public void init()
    {
        jt1 = new JTextField(40);
        jt2 = new JTextField(40);
        jt3 = new JTextField(40);
        jt3.setEditable(false);
        JPanel jp1 = new JPanel(new GridLayout(3,2));
        jp1.add(new JLabel("数值1:"));
        jp1.add(jt1);
        jp1.add(new JLabel("数值2:"));
        jp1.add(jt2);
        jp1.add(new JLabel("数值3:"));
```

```java
            jp1.add(jt3);
            b1 = new JButton("+");
            b2 = new JButton("-");
            b3 = new JButton(" * ");
            b4 = new JButton("/");
            JPanel jp2 = new JPanel(new GridLayout(1,3));
            jp2.add(b1);
            jp2.add(b2);
            jp2.add(b3);
            jp2.add(b4);
            add(jp1);
            add(jp2,BorderLayout.SOUTH);
            b1.addActionListener(this);
            b2.addActionListener(this);
            b3.addActionListener(this);
            b4.addActionListener(this);
        }
        public void actionPerformed(ActionEvent e)
        {
            Double d1 = new Double(jt1.getText());
            Double d2 = new Double(jt2.getText());
            double d = 0;
            if(e.getSource() == b1)
                d = d1+d2;
            else if(e.getSource() == b2)
                d = d1-d2;
            else if(e.getSource() == b3)
                d = d1 * d2;
            else if(e.getSource() == b4)
                d = d1/d2;
            jt3.setText((new Double(d)).toString());
        }
    }
```

html 文件设计如下：

```
<html>
<applet code="Example _ Applet13 _ 1.class" width="300" height="45">
</applet>
</html>
```

程序执行结果如图 13.1、图 13.2 所示。

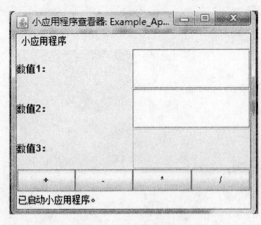
图 13.1 计算器界面

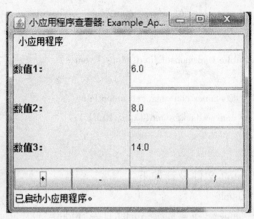
图 13.2 计算结果

【例 13.2】 Applet 程序中的鼠标事件响应。编写 Applet 程序,鼠标进出 Applet 窗口时改变背景颜色。

解析:鼠标事件响应可以通过实现 MouseListener 接口或继承 MouseAdapter 适配器实现。实现 MouseListener 接口必须覆盖其中所有的方法,但本程序只用到其中的两个,因此本程序继承 MouseAdapter 适配器。

Applet 程序已经继承了 JApplet 类,不能再继承适配器,程序中必须定义一个新的监听类继承 MouseAdapter。

程序源代码:

```
import java. awt. * ;
import java. awt. event. * ;
import javax. swing. * ;

public classMouseApplet extends JApplet
{
public void init( )
{
    MouseListener ma = new MouseListener( this) ;//创建监听对象
    this. addMouseListener( ma) ;                //注册监听对象
}
}
//MouseListener 类监听鼠标事件
class MouseListener extends MouseAdapter
{
MouseApplet ma;
MouseListener( MouseApplet ma)
{
    this. ma = ma;
}
public void mouseEntered( MouseEvent e)
{
    Container con = ma. getContentPane( ) ;
```

```
    con.setBackground(Color.red);
  }
  public void mouseExited(MouseEvent e)
  {
    Container con = ma.getContentPane();
    con.setBackground(Color.BLUE);
  }
}
```

html 文件设计如下：

```
<html>
<applet code = "MouseApplet.class" width = "300" height = "45">
</applet>
</html>
```

程序执行结果如图 13.3、图 13.4 所示。

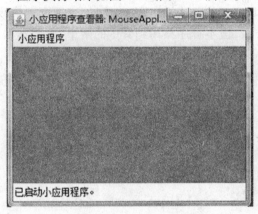

图 13.3　红色界面　　　　　　　　　　图 13.4　蓝色界面

13.2　课后习题解答

一、选择题

1~6 BABACD

二、简答题

1. Java 的 Applet 和 Java 应用程序有什么差别？

（1）运行方式不同。JavaApplet 程序不能单独运行，它必须依附于一个用 HTML 语言编写的网页并嵌入其中，通过与 Java 兼容的浏览器来控制执行。Java 应用程序是完整的程序，可以独立运行，只要有支持 Java 的虚拟机，它就可以独立运行而不需要其他文件的支持。

（2）运行工具不同。JavaApplet 必须通过网络浏览器或者 Applet 观察器才能执行。Java 应用程序通过普通的 Java 解释器就可以使其边解释边执行。

（3）程序结构不同。每个 Java 应用程序必定含有一个并且只有一个 main 方法。而 Applet 程序则没有含 main 方法的主类，这也正是 Applet 程序不能独立运行的原因。

(4) JavaApplet 程序可以直接利用浏览器或 AppletViewer 提供的图形用户界面,而 Java 应用程序则必须另外书写专用代码来营建自己的图形界面。

(5) JavaApplet 和 Java 应用程序在执行方面的主要区别表现在:Java 应用程序一般是在本地机上运行,而 JavaApplet 一般放在服务器上,它是根据本地机的请求被下载到本地机,然后才在本地机上运行。

2. 一个完整的 Applet 包括哪些基本方法?这些方法的含义分别是什么?

答:Applet 大体上有四个基本方法:

(1) init( ) 方法:这个方法用于对 applet 所需要的任何东西进行初始化,当第一次启动 applet 时,系统会自动调用这个方法。

(2) start( ) 方法:这个方法是当 Java 调用了 init 方法之后,就会自动调用的。

(3) stop( ) 方法:此方法将在系统离开 Applet 所在的 Web 页时,被调用。

(4) destroy( ) 方法:此方法会在离线时释放本对象及其相关资源,当浏览器正常关闭时,Java 会调用此方法。

3. 典型的 Applet 程序的结构是什么?

```
public class Applet extends Panel{
    //构造方法
    public Applet( )
    //部分常用的方法
    Public String getParameter(String name)
    public void init( )
    public void start( )
    public void stop( )
    public URL getCodeBase( )
    public URL getDocumentBase( )
}
```

三、程序题

1. 编制程序屏幕显示"Hello JavaApplet"。

程序代码:

```
import java.awt. * ;
import java.applet. * ;
public class HelloWorld extends Applet //继承 Applet 类,这是 Applet Java 程序的特点
{
    public void paint(Graphics g )
    {
        g.drawString("Hello JavaApplet",5,35);
    }
}
```

html 文件设计如下:

```
<html>
<applet code="HelloWorld.class" width="300" height="45">
</applet>
```

</html>

程序执行结果如图 13.5 所示。

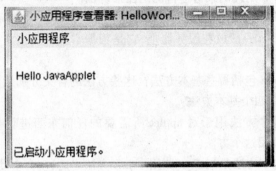

图 13.5　HelloWorld 程序执行结果

2. 编制程序屏幕显示"飞行文字",由远及近,黑色衬底,黄色字体。

```java
import java.awt.*;
import java.applet.*;
public class FlyFont extends Applet implements Runnable{
    private Image buffer;
    private Graphics gContext;
    private Font font;
    private String string;
    private Thread thread;
    private int xpos, ypos, font_size;
    public void init(){
        String param;
        buffer=createImage(getSize().width, getSize().height);
        gContext=buffer.getGraphics();
        param=getParameter("text");
        if(param==null)
            string="You Can Input The String That You Want";
        else string=param;

        font=new Font("TimesRoman", Font.BOLD, 10);

    }
    public void start(){
        if(thread==null){
            thread=  new Thread(this);
            thread.start();
        }
    }
    public void update(Graphics g){
        paint(g);
    }
```

```
public void paint(Graphics g){
    gContext.setColor(Color.black);
    gContext.fillRect(0,0,getSize().width,getSize().height);

    font=new Font("TimesRoman",Font.BOLD,font_size);
    gContext.setFont(font);
    gContext.setColor(Color.yellow);
    FontMetrics fm=gContext.getFontMetrics(font);
    int font_height=fm.getHeight();
    int w;
    int base_line=getSize().height/2+font_height/2;

    w=fm.stringWidth(string);
    w=(getSize().width-w)/2;

    gContext.drawString(string,w,base_line-=20);
    g.drawImage(buffer,0,0,this);
    font_size++;
}
public void run(){
    while(true){
        repaint();
        if(font_size >getSize().height)
            font_size=0;
        try{
            Thread.sleep(50);
        }catch (InterruptedException e){}
    }
}
}
```

html 文件设计如下:
```
<applet code=FlyFont.class width=400 height=100 VIEWASTEXT>
    <param name="text" value="飞行文字">
</applet>
```
程序执行结果如图 13.6 所示。

图 13.6　FlyFont 程序执行结果

13.3　上机实验

一、实验目的

1. 熟悉 JavaApplet 程序的构成和运行方式；
2. 熟悉 Applet 类的生命周期；
3. 熟悉 JavaApplet 传递参数的方法；
4. 学习 JavaApplet 编程方法。

二、实验内容

1. 编写一个 JavaApplet 程序，一幅图片，让图片倒映在水中。创建 Ripple 类，并继承 Applet implements Runnable 类。

2. 编写一个 JavaApplet 程序，构造一个电子相册，当鼠标选择相册一时，显示一张图片；当鼠标选择相册二时，显示另一张图片，以此类推。

三、实验要求

1. 必须编写 JavaApplet 程序；
2. 能够显示图片；
3. 能够通过传递参数控制一种或者集中图形的显示；
4. 能够通过选择控件来显示不同的图片；
5. 写出实验报告。

13.4　程序代码

1. 编写一个 JavaApplet 程序，一幅图片，让图片倒映在水中。创建 Ripple 类，并继承 Applet implements Runnable 类。

```java
import java.applet.*;
import java.awt.*;
public class Ripple extends Applet implements Runnable
{
    Thread thread=null;
    private Graphics g, refraction;
    private Image image, refimage;
    private int currentImg;
    private int imageW =0, imageH=0;
    private int ovalW =0, ovalH=0;

    private boolean    finishLoad=false;
    private final int frames=12;
    private String name="";
```

```java
    public void init( ) {
        String param;
        param = getParameter("image");
        if (param != null)
            name = param;
    }
    public void paint(Graphics g) {
        if (! finishLoad)
            return;
        if (refimage != null) {
            g.drawImage (refimage, (-currentImg * imageW), imageH, this);
            g.drawImage (refimage, ((frames-currentImg) * imageW), imageH, this);
        }
        g.drawImage (image, 0, -1, this);
    }
    public void start( ) {
        if (thread = = null) {
            thread = new Thread(this);
thread.start( );
        }
    }
    public void run( ) {
        currentImg = 0;
g = getGraphics( );
MediaTracker imageTracker = new MediaTracker(this);
String strImage;
        image = getImage(getDocumentBase( ), name);
        imageTracker.addImage(image,0);
try {
    imageTracker.waitForAll( );
    finishLoad = ! imageTracker.isErrorAny( );
}
catch (InterruptedException e) { }
        imageW = image.getWidth(this);
        System.out.println(imageW);
imageH = image.getHeight(this);
createRipple( );
        repaint( );
        while (true) {
            try {
                if (! finishLoad)
                    return;
                if (refimage != null) {
                    g.drawImage (refimage, (-currentImg * imageW), imageH, this);
```

```java
                    g.drawImage(refimage, ((frames-currentImg) * imageW), imageH, this);
                }
                g.drawImage(image, 0, -1, this);
            if(++currentImg == frames)
                currentImg = 0;
            Thread.sleep(50);
                } catch(InterruptedException e){
            stop();
                }
            }
    }
        public void createRipple(){
            Image back = createImage(imageW, imageH+1);
            Graphics offg = back.getGraphics();
            int phase = 0;
            int x, y;
            double pl;
            offg.drawImage(image, 0, 1, this);
            for(int i=0; i < (imageH >> 1); i++){
                offg.copyArea(0, i, imageW, 1, 0, imageH - i);
                offg.copyArea(0, imageH - 1 - i, imageW, 1, 0, -imageH+1+(i << 1));
                offg.copyArea(0, imageH, imageW, 1, 0, -1 - i);
            }
            refimage = createImage((frames+1) * imageW, imageH);
            refraction = refimage.getGraphics();
            refraction.drawImage(back, frames * imageW, 0, this);
            for(phase=0; phase < frames; phase++){
                pl = 2 * Math.PI * (double)phase / (double)frames;
                x = (frames - phase) * imageW;
                for(int i=0; i < imageH; i++){
                    y = (int)((imageH/14) * ((double) i+28.0)
                            * Math.sin((double)((imageH/14) * (imageH - i))/(double)(i+1)
                                + pl)/(double) imageH);
                    if(i < -y)
                refraction.copyArea(frames * imageW, i, imageW, 1,-x, 0);
                    else
                refraction.copyArea(frames * imageW, i+y,imageW, 1, -x, -y);
                }
            }
    offg.drawImage(image, 0, 1, this);
    image = back;
        }
}
```

html 文件设计如下：

```
<appletcode = Ripple. class width = 420 height = 330>
<param name = image value = "sunset. jpg">
</applet>
```

2. 编写一个 JavaApplet 程序,构造一个电子相册,当鼠标选择相册一时,显示一张图片;当鼠标选择相册二时,显示另一张图片以此类推。

```
import java. awt. * ;
import java. applet. * ;
import java. io. * ;
public class Album extends Applet
{
private Choice C1;
private String S1[ ],S2[ ];
private int totalPics;
private Image offI,img[ ];
private Graphics offG;
private MediaTracker imagetracker;
public void init( )
{
  this. setLayout( null) ;
  C1 = new Choice( ) ;
  C1. setBounds( 10 ,10 ,290 ,20) ;
  totalPics = Integer. parseInt( getParameter("TotalPic") ) ;
  System. out. println( totalPics) ;
  S1 = new String[ totalPics] ;
  S2 = new String[ totalPics] ;
  img = new Image[ totalPics] ;
  for( int i = 0; i<totalPics; i++)
  {
    S1[ i] = new String("") ;
    S2[ i] = new String("") ;
  }
  String s = new String("") ;
  imagetracker = new MediaTracker( this) ;
  for( int i = 0; i<totalPics; i++)
  {
    s = getParameter("Text"+(i+1) ) ;
    S1[ i] = s;
    System. out. println( S1[ i] ) ;
    C1. addItem( s) ;
    s = getParameter("Picture"+(i+1) ) ;
    S2[ i] = s;
    img[ i] = getImage( getDocumentBase( ) , s) ;
    imagetracker. addImage( img[ i] ,0) ;
    System. out. println( S2[ i] ) ;
```

```java
    }
    try
    {
        imagetracker.waitForID(0);
    }
    catch(InterruptedException e)
    {
    }
    add(C1);
    offI = createImage(size().width,size().height-40);
    offG = offI.getGraphics();
}
public void paint(Graphics g)
{
    g.drawImage(offI,10,40,this);
}
public boolean action(Event e , Object o)
{
    if(e.target = = C1)
    {
        String s = new String("");
        offG.setColor(this.getBackground());
        offG.fillRect(0,40,size().width,size().height-40);
        offI = img[C1.getSelectedIndex()];
        offG.drawImage(offI,0,0,this);
        repaint();
    }
    return true;
}
}
}
```

html 文件设计如下：

```html
<applet code="Album.class" width="400" height="400">
<param name="TotalPic" value="5">
<param name="Text1" value="照片一">
<param name="Text2" value="照片二">
<param name="Text3" value="照片三">
<param name="Text4" value="照片四">
<param name="Text5" value="照片五">
<param name="Picture1" value="Pic1.jpg">
<param name="Picture2" value="Pic2.jpg">
<param name="Picture3" value="Pic3.jpg">
<param name="Picture4" value="Pic4.jpg">
<param name="Picture5" value="Pic5.jpg">
</applet>
```

第14章

输入输出流

14.1 典型例题解析

【例14.1】 使用多种方式读文件内容,包括按字节、按字符等方式读取文件内容。
方式一:以字节为单位读取文件,常用于读二进制文件,如图片、声音、影像等文件。

```java
import java.io.BufferedReader;
import java.io.File;
import java.io.FileInputStream;
import java.io.FileReader;
import java.io.IOException;
import java.io.InputStream;
import java.io.InputStreamReader;
import java.io.RandomAccessFile;
import java.io.Reader;
public class ReadFromFile {
public static void readFileByBytes(String fileName) {
    File file = new File(fileName);
    InputStream in = null;
      try {
      System.out.println("以字节为单位读取文件内容,一次读一个字节");
      // 一次读一个字节
      in = new FileInputStream(file);
      int tempbyte;
      while((tempbyte = in.read()) != -1) {
         System.out.write(tempbyte);
      }
      in.close();
    } catch (IOException e) {
        e.printStackTrace();
        return;
      }
    try {
```

```java
        System.out.println("以字节为单位读取文件内容,一次读多个字节");
        //一次读多个字节
        byte[] tempbytes = new byte[100];
        int byteread = 0;
        in = new FileInputStream(fileName);
        ReadFromFile.showAvailableBytes(in);
        //读入多个字节到字节数组中,byteread 为一次读入的字节数
        while ((byteread = in.read(tempbytes)) != -1) {
            System.out.write(tempbytes, 0, byteread);
        }
    }
} catch (Exception e1) {
    e1.printStackTrace();
}
finally {
    if (in != null) {
        try {
            in.close();
        }
        catch (IOException e1) {
        }
    }
}
}
```

方式二:以字符为单位读取文件,常用于读文本、数字等类型的文件。

```java
public static void readFileByChars(String fileName) {
    File file = new File(fileName);
    Reader reader = null;
    try {
        System.out.println("以字符为单位读取文件内容,一次读一个字符:");
        // 一次读一个字符
        reader = new InputStreamReader(new FileInputStream(file));
        int tempchar;
        while ((tempchar = reader.read()) != -1) {
            //Windows 下这两个字符在一起时表示一个换行。如果这两个字符
            //   分开显示时会换行两次,将会多出很多空行,因此屏蔽掉
            if (((char)tempchar) != ' ') {
                System.out.print((char)tempchar);
            }
        }
        reader.close();
    } catch (Exception e) {
        e.printStackTrace();
    }
```

```java
try{
    System.out.println("以字符为单位读取文件内容,一次读多个字符");
    //一次读多个字符
    char[] tempchars=new char[30];
    int charread=0;
    reader=new InputStreamReader(new FileInputStream(fileName));
    //读入多个字符到字符数组中,charread 为一次读取字符数
    while((charread=reader.read(tempchars))!=-1){
        //同样屏蔽掉不显示
    if(((charread==tempchars.length)&&(tempchars[tempchars.length-1]!=' '))){
        System.out.print(tempchars);
        }
        else{
            for(int i=0;i<charread;i++){
                if(tempchars[i]==' '){
                continue;
                }
                else{
                    System.out.print(tempchars[i]);
                }
            }
        }
    }
}
catch(Exception e1){
    e1.printStackTrace();
}
finally{
    if(reader!=null){
    try{
        reader.close();
    }catch(IOException e1){
        }
    }
}
}
```

【例 14.2】 使用 RandomAccessFile,FileWrite 两种方式将内容追加到文件尾部。

```java
import java.io.FileWriter;
import java.io.IOException;
import java.io.RandomAccessFile;
//方式一:使用 RandomAccessFile 在文件尾部添加内容
public class AppendToFile{
public static void appendMethodA(String fileName, String content){
    try{
```

```java
        // 打开一个随机访问文件流,按读写方式
        RandomAccessFile randomFile = new RandomAccessFile(fileName, "rw");
        // 文件长度,字节数
        long fileLength = randomFile.length();
        //将写文件指针移到文件尾
        randomFile.seek(fileLength);
        randomFile.writeBytes(content);
        randomFile.close();
    } catch (IOException e) {
        e.printStackTrace();
    }
}

//方式二:使用 RandomAccessFile 在文件尾部添加内容
public static void appendMethodB(String fileName, String content) {
    try {
        //打开一个写文件器,构造函数中的第二个参数 true 表示以追加形式写文件
        FileWriter writer = new FileWriter(fileName, true);
        writer.write(content);
        writer.close();
    } catch (IOException e) {
        e.printStackTrace();
    }
}

public static void main(String[] args) {
    String fileName = "C:/temp/newTemp.txt";
    String content = "new append!";
    //追加文件
    AppendToFile.appendMethodA(fileName, content);
    AppendToFile.appendMethodA(fileName, "append end. ");
    //显示文件内容
    ReadFromFile.readFileByLines(fileName);
    //追加文件
    AppendToFile.appendMethodB(fileName, content);
    AppendToFile.appendMethodB(fileName, "append end. ");
    //显示文件内容
    ReadFromFile.readFileByLines(fileName);
}
}
```

【例14.3】 本例题实现文件的各种操作类,包括新建目录、新建文件、删除文件、删除文件夹、删除文件夹下的文件、复制文件、复制文件夹下所有文件、移动文件到指定目录、移动文件夹到指定目录。通过本例的学习可以掌握关于文件的大部分操作。

```java
import java.io.*;
/**
 *新建目录
```

```java
 */
public class File03{
  public void newFolder(String folderPath){
    try{
      String filePath=folderPath;
      filePath=filePath.toString();
      File myFilePath=new File(filePath);
      if(! myFilePath.exists()){
        myFilePath.mkdir();
      }
      System.out.println("新建目录操作成功执行");
    }
    catch(Exception e){
      System.out.println("新建目录操作出错");
      e.printStackTrace();
    }
  }
}
/**
 *新建文件
 */
public void newFile(String filePathAndName, String fileContent){
  try{
    String filePath=filePathAndName;
    filePath=filePath.toString();
    File myFilePath=new File(filePath);
    if (! myFilePath.exists()){
      myFilePath.createNewFile();
    }
    FileWriter resultFile=new FileWriter(myFilePath);
    PrintWriter myFile=new PrintWriter(resultFile);
    String strContent=fileContent;
    myFile.println(strContent);
    resultFile.close();
    System.out.println("新建文件操作 成功执行");
  }
  catch (Exception e){
    System.out.println("新建目录操作出错");
    e.printStackTrace();
  }
}
}
/**
 *删除文件
 */
public void delFile(String filePathAndName){
```

```java
try{
    String filePath = filePathAndName;
    filePath = filePath.toString();
    File myDelFile = new File(filePath);
    myDelFile.delete();
    System.out.println("删除文件操作 成功执行");
}
catch(Exception e){
    System.out.println("删除文件操作出错");
    e.printStackTrace();
}
}
/**
 * 删除文件夹
 */
public void delFolder(String folderPath){
try{
    delAllFile(folderPath); //删除完里面所有内容
    String filePath = folderPath;
    filePath = filePath.toString();
    File myFilePath = new File(filePath);
    if(myFilePath.delete()){ //删除空文件夹
        System.out.println("删除文件夹"+folderPath+"操作 成功执行");
    }
    else{
        System.out.println("删除文件夹"+folderPath+"操作 执行失败");
    }
}
catch(Exception e){
    System.out.println("删除文件夹操作出错");
    e.printStackTrace();
}
}
/**
 * 删除文件夹里面的所有文件
 * @param path String 文件夹路径
 */
public void delAllFile(String path){
    File file = new File(path);
    if(!file.exists()){
        return;
    }
    if(!file.isDirectory()){
        return;
```

```java
}
String[] tempList=file.list();
File temp=null;
for(int i=0; i < tempList.length; i++){
    if(path.endsWith(File.separator)){
        temp=new File(path+tempList[i]);
    }
    else{
        temp=new File(path+File.separator+tempList[i]);
    }
    if(temp.isFile()){
        temp.delete();
    }
    if(temp.isDirectory()){
        //delAllFile(path+"/"+ tempList[i]);//先删除文件夹里面的文件
        delFolder(path+ File.separatorChar+tempList[i]);//再删除空文件夹
    }
}
System.out.println("删除文件操作 成功执行");
}
/**
 * 复制单个文件
 * @param oldPath String 原文件路径 如:c:/fqf.txt
 * @param newPath String 复制后路径 如:f:/fqf.txt
 */
public void copyFile(String oldPath, String newPath){
try{
    int bytesum=0;
    int byteread=0;
    File oldfile=new File(oldPath);
    if(oldfile.exists()){
        //文件存在时
        InputStream inStream=new FileInputStream(oldPath);//读入原文件
        FileOutputStream fs=new FileOutputStream(newPath);
        byte[] buffer=new byte[1444];
        while((byteread=inStream.read(buffer)) !=-1){
            bytesum+=byteread; //字节数 文件大小
            System.out.println(bytesum);
            fs.write(buffer, 0, byteread);
        }
        inStream.close();
    }
    System.out.println("删除文件夹操作 成功执行");
}
```

```java
    catch (Exception e){
      System.out.println("复制单个文件操作出错");
      e.printStackTrace();
    }
  }
}
/**
 * 复制整个文件夹内容
 */
public void copyFolder(String oldPath, String newPath){
  try{
    (new File(newPath)).mkdirs(); //如果文件夹不存在 则建立新文件夹
    File a=new File(oldPath);
    String[] file=a.list();
    File temp=null;
    for(int i=0; i < file.length; i++){
      if(oldPath.endsWith(File.separator)){
        temp=new File(oldPath+file[i]);
      }
      else{
        temp=new File(oldPath+File.separator+file[i]);
      }
      if(temp.isFile()){
        FileInputStream input=new FileInputStream(temp);
        FileOutputStream output=new FileOutputStream(newPath+"/"+
          (temp.getName()).toString());
        byte[] b=new byte[1024 * 5];
        int len;
        while ((len=input.read(b)) !=-1){
          output.write(b, 0, len);
        }
        output.flush();
        output.close();
        input.close();
      }
      if(temp.isDirectory()){
        //如果是子文件夹
        copyFolder(oldPath+"/"+file[i],newPath+"/"+file[i]);
      }
    }
    System.out.println("复制文件夹操作 成功执行");
  }
  catch (Exception e){
    System.out.println("复制整个文件夹内容操作出错");
    e.printStackTrace();
```

```java
    }
}
/**
*移动文件到指定目录
*/
public void moveFile(String oldPath, String newPath) {
    copyFile(oldPath, newPath);
    delFile(oldPath);
}
/**
*移动文件到指定目录
*/
public void moveFolder(String oldPath, String newPath) {
    copyFolder(oldPath, newPath);
    delFolder(oldPath);
}
public static void main(String args[]) {
    String aa,bb;
    boolean exitnow=false;
    System.out.println("使用此功能请按[1] 功能一:新建目录");
    System.out.println("使用此功能请按[2] 功能二:新建文件");
    System.out.println("使用此功能请按[3] 功能三:删除文件");
    System.out.println("使用此功能请按[4] 功能四:删除文件夹");
    System.out.println("使用此功能请按[5] 功能五:删除文件夹里面的所有文件");
    System.out.println("使用此功能请按[6] 功能六:复制文件");
    System.out.println("使用此功能请按[7] 功能七:复制文件夹的所有内容");
    System.out.println("使用此功能请按[8] 功能八:移动文件到指定目录");
    System.out.println("使用此功能请按[9] 功能九:移动文件夹到指定目录");
    System.out.println("使用此功能请按[10] 退出程序");
    while(! exitnow) {
        FileOperate fo=new FileOperate();
        try {
            BufferedReader Bin=new
                BufferedReader(new InputStreamReader(System.in));
            String a=Bin.readLine();
            int b=Integer.parseInt(a);
            switch(b) {
                case 1:System.out.println("你选择了功能一 请输入目录名");
                    aa=Bin.readLine();
                    fo.newFolder(aa);
                    break;
                case 2:System.out.println("你选择了功能二 请输入文件名");
                    aa=Bin.readLine();
                    System.out.println("请输入在"+aa+"中的内容");
```

```
            bb = Bin. readLine( );
            fo. newFile( aa,bb);
            break;
      case 3:System. out. println("你选择了功能三 请输入文件名");
            aa = Bin. readLine( );
            fo. delFile( aa);
            break;
      case 4:System. out. println("你选择了功能四 请输入文件夹名");
            aa = Bin. readLine( );
            fo. delFolder( aa);
            break;
      case 5:System. out. println("你选择了功能五 请输入文件夹名");
            aa = Bin. readLine( );
            fo. delAllFile( aa);
    break;
      case 6:System. out. println("你选择了功能六 请输入文件名");
            aa = Bin. readLine( );
            System. out. println("请输入目标文件名");
            bb = Bin. readLine( );
            fo. copyFile( aa,bb);
            break;
      case 7:System. out. println("你选择了功能七 请输入源文件夹名");
            aa = Bin. readLine( );
            System. out. println("请输入目标文件夹名");
            bb = Bin. readLine( );
            fo. copyFolder( aa,bb);
            break;
      case 8:System. out. println("你选择了功能八 请输入源文件名");
            aa = Bin. readLine( );
            System. out. println("请输入目标文件夹名");
            bb = Bin. readLine( );
            fo. moveFile( aa,bb);
            break;
      case 9:System. out. println("你选择了功能九 请输入源文件夹名");
            aa = Bin. readLine( );
            System. out. println("请输入目标文件夹名");
            bb = Bin. readLine( );
            fo. moveFolder( aa,bb);
            break;
      case 10:exitnow = true;
            System. out. println("程序结束,请退出");
            break;
      default:System. out. println("输入错误. 请输入 1-10 之间的数");
      }
```

```
            System.out.println("请重新选择功能");
         }
         catch(Exception e)
         {
            System.out.println("输入错误字符或程序出错");
         }
      }
   }
}
```

14.2 课后习题解答

一、选择题

1~5 CDADB 6~10 DCBCB
11~15 ADA AB 16~20 DDBAD

二、判断题

1~5 √√×√× 6~10 √√√√√ 11~15 ×√√√√

三、简答题

1. 字节流与字符流有什么差别？

答：按处理数据的类型，流可以分为字节流与字符流，它们处理的信息的基本单位分别是字节(byte)与字符(char)。

2. 节点流与处理流有什么差别？

答：节点流(node stream)：直接与特定的地方(如磁盘、内存、设备等)相连，可以从/向一个特定的地(节点)读写数据，如文件流 FileReader。

处理流(processing stream)：是对一个已存在的流的连接和封装，通过所封装的流的功能调用实现数据读/写功能。处理流又称为过滤流，如缓冲处理流 BufferedReader。

3. 输入流与输出流各有什么方法？

答：输入流 InputStream 类最重要的方法是读数据的 read()方法。read()方法的功能是逐字节地以二进制的原始方式读入数据。另外，它还有 skip(long n)，reset()，available()，close()方法等。

输出流 OutputStream 类的重要方法是 write()，它的功能是将字节写入流中。另外，它还有 flush()及 close()方法。

4. 怎样进行文件及目录的管理？

答：Java 支持文件管理和目录管理，它们都是由专门的 java.io.File 类来实现。每个 File 类的对象表示一个磁盘文件或目录，其对象属性中包含了文件或目录的相关信息，如名称、长度、所含文件个数等，调用它的方法则可以完成对文件或目录的常用管理操作，如创建、删除等。

四、编程题

1. 编写一个程序,从命令上行接收两个实数,计算其乘积。

```java
import java.io.*;
public class Ex11_1{
public static void main(String[] args){
    String s="";
    double c=0;
    double d=0;
    try{
      BufferedReader in=new BufferedReader(
        new InputStreamReader(System.in));
      System.out.print("请输入一个数:");
      s=in.readLine();
      c=Double.parseDouble(s);
      System.out.print("请输入另一个数:");
      s=in.readLine();
      d=Double.parseDouble(s);
      System.out.println("这两个数的积为:"+(c*d));
    }catch(IOException e){}
}
}
```

2. 编写一个程序,从命令行上接收两个文件名,比较两个文件的长度及内容。

```java
public class Ex11_2{
public static void main(String[] args){
    try{
      BufferedReader in=new BufferedReader(
        new InputStreamReader(System.in));
      System.out.print("请输入一个文件名:");
      String name1=in.readLine();
      System.out.print("请输入另一个文件名:");
      String name2=in.readLine();
      File file1=new File(name1);
      File file2=new File(name2);
      long len1=file1.length();
      long len2=file2.length();
      if(len1!=len2){
        System.out.println("这两个文件的长度不一样");
        return;
      }
      String text1=readFileToEnd(file1);
      String text2=readFileToEnd(file2);
      System.out.println(text1);
```

```java
        System.out.println( text2 );
        if( text1.equals(text2) ){
            System.out.println( "这两个文件的内容相同" );
        }else{
            System.out.println( "这两个文件的内容不一样" );
        }
    }catch(IOException e){
        e.printStackTrace();
    }
}
public static String readFileToEnd( File file ){
    StringBuffer text = new StringBuffer();
    try {
        BufferedReader in = new BufferedReader(
            new FileReader(file));
        String s = in.readLine();
        while ( s != null ) {
            text.append( s+"\n" );
            s = in.readLine();
        }
        in.close();
    } catch (IOException e2) {
        e2.printStackTrace();
    }
    return text.toString();
}
}
```

3. 编写一个程序,能将一个 Java 源程序中的空行及注释去掉。

```java
import java.io.*;
public class Ex11_3 {
    public static void main (String[] args) {
        String infname = "CopyFileAddLineNumber.java";
        String outfname = "CopyFileAddLineNumber.txt";
        if( args.length >=1 ) infname = args[0];
        if( args.length >=2 ) outfname = args[1];
        try {
            File fin = new File(infname);
            File fout = new File(outfname);
            BufferedReader in = new BufferedReader(new FileReader(fin));
            PrintWriter out = new PrintWriter(new FileWriter(fout));
            int cnt = 0;// 行号
            String s = in.readLine();
            while ( s != null ) {
                cnt++;
                s = deleteComments(s);//去掉以//开始的注释
```

```java
            if( s.length() ! =0 ){
                out.println( s );//写出非空行
            }
            s=in.readLine();//读入
        }
        in.close();// 关闭缓冲读入流及文件读入流的连接
        out.close();
    } catch ( FileNotFoundException e1 ) {
        System.err.println("File not found!");
    } catch ( IOException e2 ) {
        e2.printStackTrace();
    }}
    static String deleteComments( String s ) //去掉以//开始的注释
    {
        if( s==null ) return s;
        int pos=s.indexOf( "//" );
        if( pos<0 ) return s;
        return s.substring( 0, pos );
    }
}
```

4. 编程在 D 盘上创建一个名为 test.txt 文件。

```java
import java.io.File;
import java.io.IOException ;
public class Ex11_4{
    public static void main(String args[ ]){
        File f=new File("d:\\test.txt") ;// 实例化 File 类的对象
        try{
            f.createNewFile() ;// 创建文件,根据给定的路径创建
        }catch(IOException e){
            e.printStackTrace() ;// 输出异常信息
        }
    }
}
```

5. 编程删除 D 盘中名为 text.txt 的文件。

```java
import java.io.File ;
import java.io.IOException ;
public class Ex11_5{
    public static void main(String args[ ]){
        File f=new File("d:"+File.separator+"test.txt") ;// 实例化 File 类的对象
        f.delete() ;// 删除文件
    }
}
```

6. 编程判断 D 盘上名为 test.test 文件是否存在。

```java
import java.io.File ;
```

```
import java.io.IOException ;
public class Ex11_6
{
    public static void main(String args[ ])
    {
        File f=new File("d:"+File.separator+"test.txt") ;// 实例化 File 类的对象
        if(f.exists()){// 如果文件存在则删除
            f.delete() ;// 删除文件
        }
    }
}
```

14.3　上机实验

一、实验目的与意义

1. 理解数据流的概念；
2. 理解 Java 流的层次结构；
3. 理解文件的概念；
4. 学习和掌握文件管理的编程；
5. 掌握 RandomAccessFile 类的使用。

二、实验内容

1. 编程实现文件的复制。通过字节流读取文件 C:\source.txt 中的数据读并写入到 C:\dest.txt 文件中去。
2. 编程实现文件的创建、重命名、移动和删除等文件管理功能。
3. 使用 RandomAccessFile 类统计一篇英文文章中的单词。包括实现下列功能：(1) 一共出现多少个单词？(2) 有多少互不相同的单词？(3) 统计每个单词出现的频率。

三、实验要求

1. JDK1.5 与 eclipse 开发工具；
2. 了解流的概念和相互之间的区别。

14.4　程序代码

1. 本实验功能较为简单，主要完成文件的复制，复制的方式如下：通过字节流读取文件 C:\source.txt 中的数据并写入 C:\dest.txt 文件中。通过本实验掌握 FileInputStream 和 FileOutputStream 类的常用方法。

```
import java.io.*;
public class FileCopy{
    public static void main(String[ ] args){
        // TODO Auto-generated method stub
```

```java
try {
    FileInputStream fis = new FileInputStream("c:\\source.txt");
    FileOutputStream fos = new FileOutputStream("c:\\dest.txt");
    int read = fis.read();
    while (read != -1) {
        fos.write(read);
        read = fis.read();
    }
    fis.close();
    fos.close();
}
catch (IOException e) {
    System.out.println(e);
}
}
}
```

2. 编写一个 Java 应用程序,实现文件的创建、重命名、移动和删除等文件管理功能。

```java
import java.io.*;
public class FileIO{
public static void main(String[] args) {
    FileIO fileIO = new FileIO();
}
public FileIO() {
    //创建文件
    File newFile = new File("newfile.txt");
try{
    if(newFile.exists() == false) {
        if(newFile.createNewFile() == true)
            System.out.println("Create newfile.txt success!");
        else
            System.out.println("Create newfile.txt fail!");
    }
    else
        System.out.println("file already exists!");
}catch(IOException e) {
    e.printStackTrace();
}catch(SecurityException e) {
    e.printStackTrace();
}
//重命名文件
File renameFile = new File("renamefile.txt");
newFile.renameTo(renameFile);
System.out.println(newFile.getName()+" is renamed by "+renameFile.getName());
//移动文件
```

```
File dest=new File("doc");    //生成一个目录对象
if(! dest.isDirectory()){    //测试该对象是不是目录
    System.out.println("Directory :"+dest.getName()+" is not exists!");
    try{
        if(dest.mkdir())    //根据指定的对象创建一个新的目录
            System.out.println("Directory "+dest.getName()+" is created!");
        }catch(SecurityException e){
            e.printStackTrace();
        }
    }
    File removeFile=new File(dest,renameFile.getName());
    renameFile.renameTo(removeFile);
    System.out.println(renameFile.getName()+" is removed to "+removeFile.getPath());
    //删除文件
    try{
        File delFile=new File(removeFile.getAbsolutePath());
        delFile.delete();
        System.out.println(removeFile.getName()+" in "+removeFile.getPath()+" is
            deleted!");
    }catch(SecurityException e){
        e.printStackTrace();
    }
}
}
```

程序运行结果如图 14.1 所示。

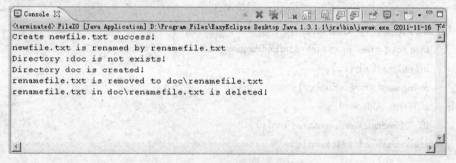

图 14.1　程序运行结果

3. 使用 RandomAccessFile 类统计一篇英文文章中的单词。包括实现下列功能：(1) 一共出现多少个单词？(2) 有多少个互不相同的单词？(3) 统计每个单词出现的频率。

```
import java.awt.*;
import java.awt.event.*;
import java.io.*;
import java.util.Vector;
class WordStatistic{
    Vector allWords,noSameWord;
    WordStatistic(){
        allWords=new Vector();
```

```java
        noSameWord=new Vector();
    }
    public void wordStatistic(File file){
        try{
        RandomAccessFile inOne=new RandomAccessFile(file,"rw");
        RandomAccessFile inTwo=new RandomAccessFile(file,"rw");
            long wordStarPosition=0,wordEndPosition=0;
            long length=inOne.length();
            int flag=1;
            int c=-1;
            for(int k=0;k<=length;k++){
                c=inOne.read();
                boolean boo=(c<='Z'&& c>='A')||(c<='z'&& c>='a');
                if(boo){
                    if(flag==1){
                        wordStarPosition=inOne.getFilePointer()-1;
                        flag=0;
                    }
                }
                else{
                    if(flag==0){
                    if(c==-1)
                        wordEndPosition=inOne.getFilePointer();
                    else
                        wordEndPosition=inOne.getFilePointer()-1;
                    inTwo.seek(wordStarPosition);
                    byte cc[]=new byte[(int)wordEndPosition-(int)wordStarPosition];
                    inTwo.readFully(cc);
                    String word=new String(cc);
                    allWords.add(word);
                    if(!(noSameWord.contains(word)))
                        noSameWord.add(word);
                    }
                    flag=1;
                }
            }
        inOne.close();
        inTwo.close();
        }catch(Exception e){}
    }
    public Vector getAllWords(){
        return allWords;
    }
    public Vector getNoSameWord(){
```

```
        return noSameWord;
    }
}
class StatisticFrame extends Frame implements ActionListener{
    WordStatistic statistic;
    TextArea showMessage;
    Button openFile;
    FileDialog openFileDialog;
    Vector allWord,noSameWord;
    public StatisticFrame(){
        statistic=new WordStatistic();
        showMessage=new TextArea();
        openFile=new Button("Open File");
        openFile.addActionListener(this);
        add(openFile,BorderLayout.NORTH);
        add(showMessage,BorderLayout.CENTER);
        openFileDialog=new FileDialog(this,"打开文件对话框",FileDialog.LOAD);
        allWord=new Vector();
        noSameWord=new Vector();
        setSize(350,300);
        setVisible(true);
        addWindowListener(new WindowAdapter(){
            public void windowsClosing(WindowEvent e){
                System.exit(0);
            }
        });
        validate();
    }
    public void actionPerformed(ActionEvent e){
        noSameWord.clear();
        allWord.clear();
        showMessage.setText(null);
        openFileDialog.setVisible(true);
        String fileName=openFileDialog.getFile();
        if(fileName!=null){
            statistic.wordStatistic(new File(fileName));
            allWord=statistic.getAllWords();
            noSameWord=statistic.getNoSameWord();
            showMessage.append("\n"+fileName+"中有"+allWord.size()+"个英文单词");
            showMessage.append("\n 其中有"+noSameWord.size()+
                    "个互不相同的英文单词");
            showMessage.append("\n 按使用频率排列:\n");
            int count[]=new int[noSameWord.size()];
            for(int i=0;i<noSameWord.size();i++){
```

```java
            String s1 = (String)noSameWord.elementAt(i);
            for(int j=0;j<allWord.size();j++){
                String s2 = (String)allWord.elementAt(j);
                count[i]++;
            }
        }
        for(int m=0;m<noSameWord.size();m++){
            for(int n=m+1;n<noSameWord.size();n++){
                if(count[n]>count[m]){
                    String temp = (String)noSameWord.elementAt(m);
                    noSameWord.setElementAt((String)noSameWord.elementAt(n),m);
                    noSameWord.setElementAt(temp,n);
                    int t=count[m];
                    count[m]=count[n];
                    count[n]=t;
                }
            }
        }
        for(int m=0;m<noSameWord.size();m++){
            showMessage.append("\n"+(String)noSameWord.elementAt(m)+":"+count[m]+"/"+allWord.size()+"="+(1.0*count[m])/allWord.size());
        }
    }
  }
 }
}
public class CountWord {
    public static void main(String[] args){
        new StatisticFrame();
    }
}
```

第 15 章

数据库编程

15.1 典型例题解析

一、数据库访问概述

1. 数据库存储的特点

相对于用文件存储数据实现 I/O,采用数据库技术存储并与应用程序交换数据具有如下特点:

(1) 相关的但不同类型的数据被集成化;
(2) 数据与程序具有相对的独立性;
(3) 可实现多个程序和用户对数据的共享;
(4) 数据的冗余度小;
(5) 避免了并发访问中数据的不一致性;
(6) 通过设置权限可对数据实施安全性保护;
(7) 有利于保证数据的完整性;
(8) 可发现存储故障并恢复到正常状态。

2. 数据库访问的概念

数据库的创建和维护、数据的访问和更新,既可以在数据库管理系统(DBMS)下进行,也可以在应用程序中实现,后者更加具有现实意义。所谓数据库访问,就是应用程序以某种方式与数据库交互,使用和更新数据库中的数据。实现访问的必要条件是操作系统、数据库管理系统和程序设计语言中有着对访问的支持,这便是数据库接口或数据库驱动程序,在 Java 中,这种接口称为 JDBC,它是由 java.sql 包中的一组类和接口组成的。

3. 数据库操作类型和操作方式

数据库访问应具有与 DBMS 对等的一系列操作,基本操作为数据的检索、修改、插入和删除;辅助操作包括数据库定义、数据转储等。本章仅介绍前一类操作。

在多数程序设计语言中,数据库访问操作可通过两种方式实现:一种是通过类库中的类所提供的 SQL 语句执行机制执行 SQL,另一种是通过类库中的类所提供的一系列方法。对某些操作而言,二者可以相互替换,对另一些操作而言,则只能使用其中的一种。而前者往往更具有普遍意义。

二、Java 数据库访问

1. ODBC 和 JDBC

JDBC(java database connectivity)是 SUN 为 Java 开发的数据库连接解决方案,它通过 JDBC-ODBC 桥接器,使用微软的 ODBC(open database connectivity)来实现与不同数据库的连接,但在编程方面要较 ODBC 更加方便。JDBC 内嵌的 SQL 为程序员提供了一个纯 Java 的数据库编程接口(由一组类和接口构成),通过它们,JDBC 实现了三个最基本的数据库访问功能:建立与数据库的连接、执行 SQL 语句和处理执行结果。

2. JDBC 的构成

java.sql 中类和接口的名称和基本功能见表 15.1。

表 15.1　java.sql 中类和接口

名称	类型	基本功能
名称	类型	基本功能
java.sql.DriverManager	类	加载 JDBC、建立与新的数据库的连接
java.sql.Connection	接口	处理与特定数据库的连接
java.sql.Statement	接口	在指定的连接中处理 SQL 语句
java.sql.PreparedStatement	子接口	处理预编译的 SQL 语句
java.sql.CallableStatement	子接口	处理数据库存储过程
java.sql.ResultSet	接口	处理数据库操作结果集

3. JDBC Driver

上述类和接口都是抽象的,在不同环境下与数据库的连接,还要靠 JDBC Driver 来实现,见表 15.2。

表 15.2　JDBC Driver

JDBC Driver 名称	连接机制和环境要求
JDBC-ODBC Bridge and ODBC driver	通过 ODBC 与数据库实现连接,要求每台客户机都装有 ODBC 驱动程序
Native-API partly-JavaDriver	将 JDBC 指令转化为 DBMS 操作形式,要求客户机装有相应的 DBMS
JDBC-Net All-JavaDriver	将 JDBC 指令转化为独立于 DBMS 的网络协议,再由服务器转化为特定的 DBMS 协议。可连接到不同的数据库
Native-protocol All-JavaDriver	将 JDBC 指令转化为网络协议,并由 DBMS 直接使用。适用于局域网

4. 通过 JDBC 访问数据库的步骤

(1)创建指定数据源的 URL。要创建与数据库的连接,首先要创建针对指定数据源的 URL(uniform resource locator),创建 URL 的一般形式如下:

String url="jdbc：odbc：数据源名称";

假设在 ODBC 管理器中设置的数据源名称为 Student,则创建语句为

String url="jdbc:odbc:Student";

数据源名称可以在 Windows 的"控制面板→管理工具→数据源(ODBC)"中针对特定的数据库指定,操作如下:添加→Microsoft Access Driver→完成,输入数据源名,选择(数据库)→确定。

(2) 加载数据库驱动程序。为了连接到具体的数据库,JDBC 必须首先加载与该数据库相应的驱动程序,代码形式如下:

Class. forName("sun. jdbc. odbc. JdbcOdbcDriver");

其中 Class 是包 java. lang 中的一个类:

public final class Class<T> extends Object implements Serializable, GenericDeclaration, Type, notatedElement

它通过调用静态方法 forName,加载 sun. jdbc. odbc 包中的 JdbcOdbcDriver 类来建立 JDBC-ODBC 桥接器。

(3) 创建数据库连接。调用 DriverManager 类中的方法 getConnection 建立与具体数据库的连接,并将该连接的引用赋给 Connection 类的对象,代码形式如下:

Connection con = DriverManager. getConnection("URL","用户名","口令");

例如,对于上述的 URL 和 Access 数据库,可以是:

Connection con = DriverManager. getConnection (url, "user", "password");

(4) 创建语句对象。声明一个 Statement 对象,通过 Connection 的对象调用方法 createStatement,并将返回值赋给这个 Statement 对象,代码形式可以是以下两种:

Statement state = con. createStatement( );

Statement state = con. createStatement(int resultSetType, int resultSetConcurrency);

其中变量见表 15.3。

表 15.3 resultSetType 与 resultSetConcurrency

变量	取值	意义
resultSetType	ResultSet. TYPE _ FORWARD _ ONLY(默认)	游标只能向下滚动
	ResultSet. TYPE _ SCROLL _ INSENSITIVE	上下滚动,不同步更新
	ResultSet. TYPE _ SCROLL _ SENSITIVE	上下滚动,同步更新
resultSetConcurrency	ResultSet. CONCUR _ READ _ ONLY(默认)	不允许更新数据库
	ResultSet. CONCUR _ UPDATABLE	允许更新数据库

(5) 执行 SQL 查询。有了 Statement 对象后,就可以通过它调用 Statement 的方法来对数据库进行各种操作,操作命令由 SQL 语句给出,查询结果保存在 ResultSet 类(查询结果数据集)的对象中返回,假设一个 SQL 语句定义在字符串 sql 中:

String sql = "SELECT * FROM 学生";

则:

ResultSet result = state. executeUpdate(sql);

将从学生表中查询出所有数据并保存在对象 result 中。

(6) 处理查询结果。利用 resultSet 类提供的一系列方法,可以对进行各种操作,见表 15.4。

表15.4 查询数据集结果

方法原型	功能说明
String getString(int columnIndex) throws SQLException	获取字符串
int getInt(int columnIndex) throws SQLException	获取整型数
double getDouble(int columnIndex) throws SQLException	获取双精度数
int getRow() throws SQLException	获取当前行编号
booleanlast() throws SQLException	移动游标到末行
void beforeFirst() throws SQLException	移动游标到首行之前
booleannext() throws SQLException	移动游标到下一行
voidclose() throws SQLException	关闭结果集

（7）更新、添加和删除记录。使用 UPDATE，INSERT INTO，DELETE FROM 等 SQL 语句，也可对数据库的某个表进行记录的更新、插入和删除操作。此时代替声明 ResultSet 对象并通过 Statement 对象调用 executeQuery 方法执行 SQL 语句并处理查询结果，是通过 Statement 对象调用 executeUpdate 方法来更新数据库，例如：

sql="UPDATE 学生 SET 性别='男' WHERE 学生 ID=1";
state.executeUpdate(sql);

三、数据库访问实例

1. Student 数据库

我们尝试运用上面的步骤来访问一个 Access 数据库 Student，其中有 4 个表，E-R 图如图 15.1 所示。

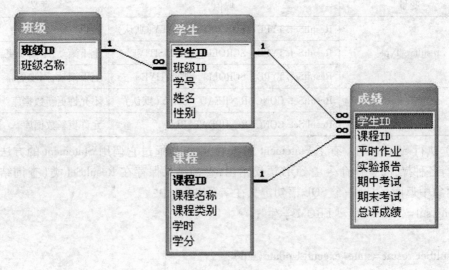

图 15.1 E-R 图

2. 设计查询

该数据库中存放着 061431 和 061432 两个班学生的基本资料和"C++程序设计"、"程序设计基础"两门课的成绩数据。现要从该数据库中查询出两个班所有同学"C++程序设计"（课

程 ID=2)的实验报告和期末考试成绩,设计 SQL 语句如下:

SELECT 学生.学号,学生.姓名,成绩.实验报告,成绩.期末考试,成绩.总评成绩
FROM 学生 INNER JOIN(课程 INNER JOIN 成绩 ON 课程.课程 ID=成绩.课程 ID)ON 学生.学生 ID=成绩.学生 ID
WHERE((课程.课程 ID)=2);

3. 在 Java 中执行查询

```java
import java.sql.*;
public class UseJdbcQuery {
    public static void main(String[] args) {
        String url = "jdbc:odbc:Student";
        String sql = "SELECT 学生.学号,学生.姓名,成绩.实验报告,成绩.期末考试 FROM 学生 INNER JOIN (课程 INNER JOIN 成绩 ON 课程.课程 ID=成绩.课程 ID) ON 学生.学生 ID=成绩.学生 ID WHERE (((课程.课程 ID)=2));";
        try {
            Class.forName("sun.jdbc.odbc.JdbcOdbcDriver");
            Connection con = DriverManager.getConnection(url, "user", "password");
            Statement state = con.createStatement();
            ResultSet result = state.executeQuery(sql);
            while (result.next()) {
                String ID = result.getString("学号");
                String name = result.getString("姓名");
                double experiment = result.getDouble("实验报告");
                double exam = result.getDouble("期末考试");
                System.out.println(result.getRow() + "\t" + ID + "\t"
                    + name + "\t" + experiment + "\t" + exam);
            }
            result.close();
            state.close();
            con.close();
        }
        catch(java.lang.Exception ex) {
            ex.printStackTrace();
        }
    }
}
```

4. 在 Java 中更新、添加和删除记录

```java
import java.sql.*;
public class UseJdbcUpdate {
    public static void main(String[] args) {
        String url, sql;
        url = "jdbc:odbc:Student";
        try {
            Class.forName("sun.jdbc.odbc.JdbcOdbcDriver");
```

```java
            Connection con=DriverManager.getConnection(url,"user","password");
            Statement state=con.createStatement();
            sql="UPDATE 成绩 SET 期末考试=100 WHERE 学生ID=1 AND
              课程ID=1";
            state.executeUpdate(sql);
            sql="INSERT INTO 学生 VALUES(88,2,'06143245','张三','男')";
            state.executeUpdate(sql);
            sql="INSERT INTO 学生 VALUES(89,2,'06143146','李四','女')";
            state.executeUpdate(sql);
            sql="DELETE FROM 学生 WHERE 学号='06143146'";
            state.executeUpdate(sql);
            state.close();
            con.close();
        }
        catch(java.lang.Exception ex){
            ex.printStackTrace();
        }
    }
}
```

15.2 课后习题解答

一、选择题

1~6 ADDDDA

二、填空题

1. 数据库 2. JDBC API 3. JDBC DriverAPI
4. java.Sql.*包 5. java.sql 6. 关闭连接

三、简答题

1. 试述JDBC提供了哪几种连接数据库的方法。

答:JDBC连接数据库的方法取决于JDBC驱动程序类型,Java定义了4种JDBC驱动程序类型:

(1) JDBC-ODBC桥驱动程序。JDBC-ODBC桥接器负责将JDBC转换为ODBC,用JdbcOdbc.Class和一个用于访问ODBC驱动程序的本地库实现的。这类驱动程序必须在服务器端安装好ODBC驱动程序,然后通过JDBC-ODBC的调用方法,进而通过ODBC来存取数据库。

(2) Java到本地API。这种类型的驱动程序是部分使用Java语言编写和部分使用本机代码编写的驱动程序,这类驱动程序也必须在服务器端安装好特定的驱动程序,如ODBC驱动程序,然后通过桥接器的转换,把JavaAPI调用转换成特定驱动程序的调用方法,进而操作数据库。

(3)网络协议搭配的 Java 驱动程序。这种驱动程序将 JDBC 转换为与 DBMS 无关的网络协议,这种协议又被某个服务器转换为一种 DBMS 协议。这种网络服务器中间件能够将它的纯 Java 客户机连接到多种不同的数据库上。所用的具体协议取决于提供者。

(4)本地协议纯 Java 驱动程序。这种类型的驱动程序将 JDBC 访问请求直接转换为特定数据库系统协议。不但无须在使用者计算机上安装任何额外的驱动程序,也不需要在服务器端安装任何中间程序,所有对数据库的操作,都直接由驱动程序来完成。

2. SQL 语言包括哪几种基本语句来完成数据库的基本操作?

答:SQL 语言包括以下 6 种基本语句来完成数据库的基本操作:

(1)select 语句:用来对数据库进行查询并返回符合用户查询标准的数据结果。

(2)create table 语句:用来建立新的数据表。

(3)insert 语句:向数据表中插入或添加新的数据行。

(4)update 语句:更新或修改符合规定条件的记录。

(5)delete 语句:删除数据表中的行或记录。

(6)drop table 语句:删除某个数据表以及该表中的所有记录。

3. Statement 接口的作用是什么?

答:Statement 接口用于执行静态 SQL 语句并返回它所生成结果的对象。在默认情况下,同一时间每个 Statement 对象只能打开一个 ResultSet 对象。因此,如果读取一个 ResultSet 对象与读取另一个交叉,则这两个对象必须是由不同的 Statement 对象生成的。如果存在某个语句打开当前 ResultSet 对象,则 Statement 接口中的所有执行方法都会隐式关闭它。

4. ExecuteQuery( )的作用是什么?

答:ExecuteQuery( )方法执行给定的 SQL 语句,返回单个 ResultSet 对象。发送给数据库的 SQL 语句,通常为静态 SQL SELECT 语句,返回包含给定查询所生成数据的 ResultSet 对象。

5. 试述 DriverManager 对象建立数据库连接所用的几种不同的方法。

答:DriverManager 对象建立数据库连接的方法有以下几种:

(1)static Connection getConnection(String url):使用指定的数据库 URL 创建一个连接。

(2)static Connection getConnection(String url, Properties info):使用指定的数据库 URL 和相关信息(用户名、用户密码等属性列表)来创建一个连接,使 DriverManager 从注册的 JDBC 驱动程序中选择一个适当的驱动程序。

(3) static Connection getConnection(String url, String user, String password):使用指定的数据库 URL、用户名和用户密码创建一个连接,使 DriverManager 从注册的 JDBC 驱动程序中选择一个适当的驱动程序。

(4) static Driver getDriver(String url):定位在给定 URL 下的驱动程序,让 DriverManager 从注册的 JDBC 驱动程序中选择一个适当的驱动程序。

四、编程题

1. 有 3 个表:Employee 职工(工号,姓名,性别,年龄,部门)(num, name, sex, age, departmentno);Wage 工资(编号,工资金额)(No, amount);Attend 出勤(工号,工资编号,出勤率)(num, No, attendance)请根据要求,编写相应的 SQL 语句:

(1)写一个 SQL 语句,查询工资金额为 8 000 的职工工号和姓名。

(2)写一个 SQL 语句,查询职工张三的出勤率。

(3)写一个SQL语句,查询3次出勤率为0的职工姓名和工号。

(4)写一个SQL语句,查询出勤率为10并且工资金额小于2 500的职工信息。

答:(1)select eml. num,eml. name from Employee eml where (select count(*) from Attend where num=eml. num and No=(select No from Wage where amount='8000'));

(2)select amount from Wage where No in (select No from Attend where num=(select num from Employee where name='张三'));

(3)select eml. num, eml. name from Employee eml where (select count(*) from Attend where num=eml. num and attendance=0)=3;

(4)select * from Employee eml where (select No from Attend where num=eml. num and attendance='10')=(select No from Wage where amount<2500);

2. 请编写访问MySQL数据库的JDBC连接代码,查询数据库中user表的全部内容,并打印出来。

```
import java. sql. Connection;
import java. sql. DriverManager;
import java. sql. ResultSet;
import java. sql. SQLException;
import java. sql. Statement;
public class DbConn {
private static DataSource ds=null;
private static Connection conn=null;
public static Connection getConn _ jdbc() {    // 使用JDBC连接数据库
  try {
    String url="jdbc:mysql://localhost:3306/bbsdb";
    String username="root";
    String password="pla";
    Class. forName("com. mysql. jdbc. Driver"). newInstance();
    conn=DriverManager. getConnection(url, username, password);
    return conn;
  }catch (Exception e) {
    System. err. println("数据库连接异常:"+e. getMessage());
    return null;
  }
}
public void CloseConn() {            // 关闭数据库连接
  try {
    conn. close();
  }catch (Exception e) {
    System. err. println("数据库连接关闭异常:"+e. getMessage());
  }
}
public static void main(String[ ] a) {//测试数据库连接
  Connection conn;
  DbConn dc=new DbConn();
```

```
conn=dc.getConn_jdbc();
try{
   Statement stmt=conn.createStatement();
   String strSql="select * from user";
   ResultSet rs=stmt.executeQuery(strSql);
   while(rs.next()){
      System.out.println("name:"+rs.getString("name"));
   }
}catch(SQLException e){
   e.printStackTrace();
}finally{
   dc.CloseConn();// 注意,必须在最后关闭数据库连接,否则将严重影响系统性能
   }
  }
}
```

15.3 上 机 实 验

一、实验目的与意义

1. 掌握数据库的连接；
2. 掌握数据库查询操作；
3. 掌握数据库的更新、添加以及删除操作。

二、实验内容

编写一个学生信息处理小软件,要求如下：
(1)准备一个 Access 数据库文件,内有一张学生信息表,字段自定；
(2)编写访问数据库的类,对数据库实现查询、更新、添加、删除操作；
(3)建立用户访问界面,如图 15.2 所示,对学生表进行正确的访问。

图 15.2 运行图

三、实验要求

1. JDK1.6 以及 Eclipse 开发环境；
2. 参考相关书籍，能够熟练地安装和使用 JDK 开发工具。

15.4 程序代码

```java
File DataAccess.java
importjava.sql.Connection;
importjava.sql.DriverManager;
importjava.sql.ResultSet;
importjava.sql.SQLException;
importjava.sql.Statement;
public class DataAccess{
Connection con;
Statement sql;
ResultSet rs;
staticboolean t;
staticint rows=0,cols=0;
publicDataAccess(){
  try{
    Class.forName("sun.jdbc.odbc.JdbcOdbcDriver");
  }catch(ClassNotFoundException e){
    e.printStackTrace();
  }
}
publicbooleanopenCon(){
  try{
  con=DriverManager.getConnection("jdbc:odbc:jsj","","");//jdbc:odbc:后加数据库名
    return true;
  }catch(SQLException e){
    e.printStackTrace();
    return false;
  }
}
publicintgetrows(){
  return rows;
}
public Object[][] getQuery(String SQL){
  rows=0;
  cols=0;
  String result[][]=null;
  inti=0,j;
```

```java
        try{sql=con.createStatement(ResultSet.TYPE_SCROLL_SENSITIVE,
                ResultSet.CONCUR_UPDATABLE);
    ResultSet fields=con.getMetaData().getColumns(null,null,"10rj2",null);
        //"10rj2"为表明
        //执行SQL语句
        if(SQL!=""){
            sql.execute(SQL);
        }
        rs=sql.executeQuery("SELECT * FROM 10rj2");
        while(fields.next()) cols++;
        while(rs.next()) rows++;
        rs.beforeFirst();
        result=new String[rows][cols];
        while(rs.next()){
            for(j=0;j<cols;j++){
                result[i][j]=rs.getString(j+1);
            }
            i++;
        }
        con.close();
        return result;
        }catch (SQLException e){
            // TODO Auto-generated catch block
            System.out.println("get wrong!");
        }
        return result;
    }
}
File win.java
importjava.awt.BorderLayout;
importjava.awt.event.ActionEvent;
importjava.awt.event.ActionListener;
importjavax.swing.Box;
importjavax.swing.JButton;
importjavax.swing.JFrame;
importjavax.swing.JLabel;
importjavax.swing.JOptionPane;
importjavax.swing.JPanel;
importjavax.swing.JScrollPane;
importjavax.swing.JTable;
importjavax.swing.JTextArea;
importjavax.swing.JTextField;
public class win extends JFrame implements ActionListener{
JTable table;
```

```java
int i=0;
int row;
DataAccess data;
Object [][] in;
String []title={"num","name","birth","gender"};
Box basebox,baseinfor,inforbox1,inforbox2,buttonbox,消息盒;
Box nunpan,namepan,birthpan,genderpan;
JLabel 学号,姓名,生日,性别;
JLabel 消息;
JTextField num,name,birth,gender;
JButton 上一个,下一个,更新,删除,添加;
public win(){
    setTitle("学生信息数据库");
    setBounds(200,200,420,200);
    init();
    setVisible(true);
    setDefaultCloseOperation(JFrame.EXIT_ON_CLOSE);
}
public void init(){
    //创建盒式布局
    basebox=Box.createVerticalBox();
    baseinfor=Box.createHorizontalBox();
    inforbox1=Box.createVerticalBox();
    inforbox2=Box.createVerticalBox();
    buttonbox=Box.createHorizontalBox();
    nunpan=Box.createHorizontalBox();
    namepan=Box.createHorizontalBox();
    birthpan=Box.createHorizontalBox();
    genderpan=Box.createHorizontalBox();
    消息盒=Box.createHorizontalBox();
    //创建文本框
    学号=new JLabel("学号");
    姓名=new JLabel("姓名");
    生日=new JLabel("生日");
    性别=new JLabel("性别");
    //把从数据库中获取的信息分配到各各文本框中
    data=new DataAccess();
    data.openCon();
    in=data.getQuery("");
    row=data.getrows();
    num=new JTextField(""+in[0][0]);
    name=new JTextField(""+in[0][1]);
    birth=new JTextField(""+in[0][2]);
    gender=new JTextField(""+in[0][3]);
```

```java
//创建按钮
下一个 = new JButton("下一个");
下一个.addActionListener(this);
上一个 = new JButton("上一个");
上一个.addActionListener(this);
更新 = new JButton("更新");
更新.addActionListener(this);
删除 = new JButton("删除");
删除.addActionListener(this);
添加 = new JButton("添加");
添加.addActionListener(this);

//在窗口上添加控件
nunpan.add(学号);
nunpan.add(num);
inforbox1.add(nunpan);
inforbox1.add(Box.createVerticalStrut(10));
namepan.add(姓名);
namepan.add(name);
inforbox1.add(namepan);
birthpan.add(生日);
birthpan.add(birth);
inforbox2.add(birthpan);
inforbox2.add(Box.createVerticalStrut(10));
genderpan.add(性别);
genderpan.add(gender);
inforbox2.add(genderpan);
baseinfor.add(inforbox1);
baseinfor.add(Box.createHorizontalStrut(15));
baseinfor.add(inforbox2);
basebox.add(baseinfor);
basebox.add(Box.createVerticalStrut(30));
buttonbox.add(下一个);
buttonbox.add(Box.createHorizontalStrut(10));
buttonbox.add(上一个);
buttonbox.add(Box.createHorizontalStrut(10));
buttonbox.add(更新);
buttonbox.add(Box.createHorizontalStrut(10));
buttonbox.add(删除);
buttonbox.add(Box.createHorizontalStrut(10));
buttonbox.add(添加);
basebox.add(buttonbox);
basebox.add(Box.createVerticalStrut(10));
消息 = new JLabel("该表共有学生"+row+"条,现在是第"+(i+1)+"号记录");
```

```java
        消息盒.add(消息);
        basebox.add(消息盒);
        basebox.add(Box.createVerticalStrut(20));
    add(basebox);
}
public void actionPerformed(ActionEvent e){
    // TODO Auto-generated method stub
    if(e.getSource()==下一个){
        if((i!=row-1)){
            i++;
            num.setText(""+in[i][0]);
            name.setText(""+in[i][1]);
            birth.setText(""+in[i][2]);
            gender.setText(""+in[i][3]);
            消息.setText("该表共有学生"+row+"条,现在是第"+(i+1)+"号记录");
        }
        else
            JOptionPane.showMessageDialog(this,"没有下一号学生了!","消息
                对话框",JOptionPane.INFORMATION_MESSAGE);//弹出消息框
    }
    else if(e.getSource()==上一个){
        if(i!=0){
            i--;
            num.setText(""+in[i][0]);
            name.setText(""+in[i][1]);
            birth.setText(""+in[i][2]);
            gender.setText(""+in[i][3]);
            消息.setText("该表共有学生"+row+"条,现在是第"+(i+1)+"号记录");
        }
        else
            JOptionPane.showMessageDialog(this,"没有上一号学生了!",
                "消息对话框",JOptionPane.INFORMATION_MESSAGE);//弹出消息框
    }
    else if(e.getSource()==更新){
        String t1=num.getText();
        String t2=name.getText();
        String t3=birth.getText();
        String t4=gender.getText();
        data=new DataAccess();
        data.openCon();
        in=data.getQuery("UPDATE 10rj2 SET gender='"+t4+"',stuName='"+ t2+"',
            birthday='"+t3+"' WHERE stuID="+t1);
    }
    else if(e.getSource()==删除){
```

```
            String d1 = num. getText( );
            data = new DataAccess( );
            data. openCon( );
            in = data. getQuery("DELETE FROM 10rj2 WHERE stuID ="+d1);
            row = data. getrows( );
            消息. setText("该表共有学生"+row+"条,现在是第"+(i+1)+"号记录");
            if( i = = row) {
                num. setText(""+in[i-1][0]);
                name. setText(""+in[i-1][1]);
                birth. setText(""+in[i-1][2]);
                gender. setText(""+in[i-1][3]);
                i--;
                消息. setText("该表共有学生"+row+"条,现在是第"+(i+1)+"号记录");
            }
            else{
                num. setText(""+in[i][0]);
                name. setText(""+in[i][1]);
                birth. setText(""+in[i][2]);
                gender. setText(""+in[i][3]);
            }
        }
        else if( e. getSource( ) = = 添加) {
            String a1 = num. getText( );
            String a2 = name. getText( );
            String a3 = birth. getText( );
            String a4 = gender. getText( );
            data = new DataAccess( );
            data. openCon( );
            in = data. getQuery("INSERTINTO 10rj2 VALUES("+a1+",'"+a2+"','"+a3+"','"+a4+"')");
            row = data. getrows( );
            if( (i+2) = = row) {
                i++;
                消息. setText("该表共有学生"+row+"条,现在是第"+(i+1)+"号记录");}
            else
                消息. setText("该表共有学生"+row+"条,现在是第"+(i+1)+"号记录");
        }
    }
    public static void main( String[ ] args) {
        win window = new win( );
    }
}
```

第16章

网络程序设计

16.1 典型例题解析

【例 16.1】 生成一个 URL 对象,并获取它的各个属性。

```java
import java.net.*;
import java.io.*;
public class ParseURL{
    public static void main (String [] args) throws Exception{
        URL Aurl=new URL("http://java.sun.com:80/docs/books/");
        URL tuto=new URL(Aurl,"tutorial.intro.html#DOWNLOADING");
        System.out.println("protocol ="+ tuto.getProtocol());
        System.out.println("host ="+ tuto.getHost());
        System.out.println("filename ="+ tuto.getFile());
        System.out.println("port ="+ tuto.getPort());
        System.out.println("ref ="+tuto.getRef());
        System.out.println("query ="+tuto.getQuery());
        System.out.println("path ="+tuto.getPath());
        System.out.println("UserInfo ="+tuto.getUserInfo());
        System.out.println("Authority ="+tuto.getAuthority());
    }
}
```

程序执行结果如图 16.1 所示。

```
protocol=http
host =java.sun.com
filename=/docs/books/tutorial.intro.html
port=80
ref=DOWNLOADING
query=null
path=/docs/books/tutorial.intro.html
UserInfo=null
Authority=java.sun.com:80
```

图 16.1 程序运行结果

【例 16.2】 从 URL 读取 WWW 网络资源。当我们得到一个 URL 对象后,就可以通过它读取指定的 WWW 资源,本例题通过 URL 对象的 getContent() 方法下载一图象文件并在 Frame 中显示。

```java
import java.net.*;
import java.io.*;
import java.awt.*;
import java.awt.image.*;
import java.awt.event.*;
public class URLTest extends Frame{
    private Image img;
    public void paint(Graphics g){
        g.drawImage(img,20,20,this);
    }
    public void processWindowEvent(WindowEvent e){
        super.processWindowEvent(e);
        if(e.getID()==WindowEvent.WINDOW_CLOSING)
            System.exit(0);
    }
    public static void main(String args[]) throws MalformedURLException,IOException{
        if(args.length!=1){ //检查用户是否提供了程序所需的命令行参数
            System.out.println("Usage:javaURLTest2<imageurl>");
            System.exit(-1);
        }
        URL url=new URL(args[0]); //创建 URL 对象
        URLTest urlt=new URLTest();
        urlt.img=urlt.createImage((ImageProducer)url.getContent());
        urlt.enableEvents(AWTEvent.WINDOW_EVENT_MASK);
        urlt.setSize(600,400);
        urlt.setVisible(true);
    }
}
```

【例 16.3】 本题是一个典型的用 Socket 实现的客户和服务器交互的 C/S 结构程序,仔细阅读该程序,会对前面所讨论的知识点有更深刻的认识,程序已在注释中作了详细的说明。

1. 客户端程序

```java
import java.io.*;
import java.net.*;
public class TalkClient{
    public static void main(String args[]){
        try{
            Socket socket=new Socket("127.0.0.1",4700);
            //向本机的 4700 端口发出客户请求
            BufferedReader sin=new BufferedReader(new
                InputStreamReader(System.in));
```

```java
        //由系统标准输入设备构造 BufferedReader 对象
        PrintWriter os = new PrintWriter(socket.getOutputStream());
        //由 Socket 对象得到输出流,并构造 PrintWriter 对象
        BufferedReader is = new BufferedReader(new
        InputStreamReader(socket.getInputStream()));
        //由 Socket 对象得到输入流,并构造相应的 BufferedReader 对象
        String readline;
        readline = sin.readLine(); //从系统标准输入读入一字符串
        while(! readline.equals("bye")){
            //若从标准输入读入的字符串为"bye",则停止循环
            os.println(readline);
            //将从系统标准输入读入的字符串输出到 Server
            os.flush();
            //刷新输出流,使 Server 马上收到该字符串
            System.out.println("Client:"+readline);
            //在系统标准输出上打印读入的字符串
            System.out.println("Server:"+is.readLine());
            //从 Server 读入一字符串,并打印到标准输出上
            readline = sin.readLine(); //从系统标准输入读入一字符串
        } //继续循环
        os.close(); //关闭 Socket 输出流
        is.close(); //关闭 Socket 输入流
        socket.close(); //关闭 Socket
    }catch(Exception e){
        System.out.println("Error"+e); //出错,则打印出错信息
    }
  }
}
```

2. 服务器端程序

```java
import java.io.*;
import java.net.*;
import java.applet.Applet;
public class TalkServer{
    public static void main(String args[]){
        try{
            ServerSocket server=null;
            try{
                server=new ServerSocket(4700);
                //创建一个 ServerSocket 在端口 4700 监听客户请求
            }catch(Exception e){
                System.out.println("can not listen to:"+e);
                //出错,打印出错信息
            }
            Socket socket=null;
```

```java
        try{
            socket=server.accept();
            //使用accept()阻塞等待客户请求,有客户
            //请求到来则产生一个Socket对象,并继续执行
        }catch(Exception e){
            System.out.println("Error."+e);
            //出错,打印出错信息
        }
        String line;
        BufferedReader is=new BufferedReader(
            new InputStreamReader(socket.getInputStream()));
        //由Socket对象得到输入流,并构造相应的BufferedReader对象
        PrintWriter os=newPrintWriter(socket.getOutputStream());
        //由Socket对象得到输出流,并构造PrintWriter对象
        BufferedReader sin=new BufferedReader(
            new InputStreamReader(System.in));
        //由系统标准输入设备构造BufferedReader对象
        System.out.println("Client:"+is.readLine());
        //在标准输出上打印从客户端读入的字符串
        line=sin.readLine();
        //从标准输入读入一字符串
        while(!line.equals("bye")){
        //如果该字符串为"bye",则停止循环
            os.println(line);
            //向客户端输出该字符串
            os.flush();
            //刷新输出流,使Client马上收到该字符串
            System.out.println("Server:"+line);
            //在系统标准输出上打印读入的字符串
            System.out.println("Client:"+is.readLine());
            //从Client读入一字符串,并打印到标准输出上
            line=sin.readLine();
            //从系统标准输入读入一字符串
        }   //继续循环
        os.close();   //关闭Socket输出流
        is.close();   //关闭Socket输入流
        socket.close();   //关闭Socket
        server.close();   //关闭ServerSocket
    }catch(Exception e){
        System.out.println("Error:"+e);
        //出错,打印出错信息
    }
  }
}
```

16.2 课后习题解答

一、选择题

1~5 ACACD　6~10 BADBB
11~15 CDBDC　16~20 BADBC

二、填空题

1. InetAddress　2. ServerSocket 类　Socket 类　3. MulticastSocket　4. UDP 协议　TCP 协议　UDP 协议　5. openStream( )　InputStream openStream( )　6. alformedURLException　try…catch 语句

三、判断题

1~5 √×√√×　6~10 ×√××√

四、简答题

1. 简答使用 java Socket 创建客户端 Socket 过程的主要程序语句。

答：try{
　Socket socket=new Socket(目标主机,目标端口);
　}catch(IOException e){
　System. out. println("Error:"+e);
　}

2. 简答使用 java ServerSocket 创建服务器端 ServerSocket 过程的主要程序语句。

答：ServerSocket server=null;
　try{
　server=new ServerSocket(监听端口);
　}catch(IOException e){
　System. out. println("can not listen to :"+e);
　}
Socket socket=null;
try{
　socket=server. accept( );
　　}catch(IOException e){
System. out. println("Error:"+e);
　}

3. 写出一种使用 Java 流式套接式编程时,创建双方通信通道的语句。

答：PrintStream os=newPrintStream(new BufferedOutputStreem(socket. getOutputStream( )));
　DataInputStream is=new DataInputStream(socket. getInputStream( ));
　PrintWriter out=new PrintWriter(socket. getOutStream( ),true);
　BufferedReaderin=newButfferedReader(new InputSteramReader(Socket. getInputStream( )));

4. 简述建立功能齐全的 Socket 基本的步骤。

答:(1)创建 Socket;
(2)打开连接到 Socket 的输入/输出流;
(3)按照一定的协议对 Socket 进行读/写操作;
(4)关闭 Socket。

5. 简述基于 TCP 及 UDP 套接字通信的主要区别。

答:TCP,可靠,传输大小无限制,但是需要连接建立时间,差错控制开销大。
UDP,不可靠,差错控制开销较小,传输大小限制在 64 K 以下,不需要建立连接。

五、编程题

1. 使用 URL 类的四种构造方法创建一个 URL 对象,URL 地址为 www.baidu.com。
(1) public URL (String spec);
   URL urlBase = new URL("http:// www.baidu.com/")
(2) public URL(URL context, String spec);
   URL net263 = new URL ("http:// www.baidu.com/");
   URL index263 = new URL(net263, "index.html")
(3) public URL(String protocol, String host, String file);
   new URL("http", "www.baidu.com", "/index.html");
(4) public URL(String protocol, String host, int port, String file);
   URL gamelan = new URL("http", "www.baidu.com", 80, "Pages/index.html");

2. 请编写 Java 程序,访问 http:// www.baidu.com 所在的主页文件。

```java
public class URLReader{
    public static void main(String[ ] args) throws Exception{
        URL tirc = new URL("http:/ www.baidu.com /");
        BufferedReader in = new BufferedReader( new
            InputStreamReader(tirc.openStream()));
        String inputLine;
        while ((inputLine = in.readLine()) ! = null)
        System.out.println(inputLine);
        in.close();
    }
}
```

3. 编程类似 ping 的程序,并测试连接效果。

```java
import java.net. * ;
import java.io. * ;
class Ping{
    public static void main(String[ ] args) {
        System.out.println("请输入 ping 主机名称或地址:\n");
        try{
            String host = keyreadline();
            InetAddress ip = InetAddress.getByName(host);
            long t1 = System.currentTimeMillis();
            if(ip.isReachable(5000)){
                long t2 = System.currentTimeMillis();
```

```
            System.out.println("\nReply from "+ip.getHostAddress()+"time<="+(t2-t1)+"ms");
            System.out.println();
        }
        else{
            System.out.println("Request timed out.");
        }}
        catch(IOException e){ System.out.println("Request timed out."); }
    }
    private static String keyreadline(){
        BufferedReader br=new BufferedReader(new
            InputStreamReader(System.in));
        String str=null;
        try{
            str=br.readLine();
            } catch(IOException e){
            e.printStackTrace();}
        return str;
    }}
```

4. 创建一个服务器,用它请求用户输入密码,然后打开一个文件,并将文件通过网络连接传送出去。创建一个同该服务器连接的客户,为其分配适当的密码,然后捕获和保存文件。在自己的机器上用 localhost(通过调用 InetAddress.getByName(null)生成本地 IP 地址 127.0.0.1)测试这两个程序。

```
import java.net.*;
import java.io.*;
public class TestClient{
public static void main(String args[]){
    try{
        Socket s1=new Socket("127.0.0.1",8888);
        InputStream is=s1.getInputStream();
        DataInputStream dis=new DataInputStream(is);
        System.out.println(dis.readUTF());
        dis.close();
        s1.close();
    } catch(ConnectException connExc){
        System.err.println("服务器连接失败!");
    } catch(IOException e){
    }}
}
//服务器端
import java.net.*;
import java.io.*;
public class TestServer{
public static void main(String args[]){
```

```
      ServerSocket s = null;
      try {
        s = new ServerSocket(8888);
      } catch (IOException e) {}
      while (true) {
        try {
          Socket s1 = s.accept();
          OutputStream os = s1.getOutputStream();
          DataOutputStream dos = new DataOutputStream(os);
          dos.writeUTF("Hello,bye-bye!");
          dos.close();
          s1.close();
        } catch (IOException e) {}
      }
    }
}
```

16.3 上机实验

一、实验目的与意义

1. 理解网络编程的基本概念和原理;
2. 理解 URL 的概念,掌握 URL 类的功能和编程方法;
3. 通过 Socket 编程,掌握网络应用程序开发的基本方法和步骤。

二、实验内容

1. 使用 URL 类获取网络信息和资源,包括获取 URL 信息和通过 URL 获取网站的主页文件;
2. 利用 URLConnection 读取网络图片并保存到本地磁盘上;
3. 使用 Socket 类分别实现服务器端程序和客户端程序。

三、实验要求

1. JDK1.5 与 eclipse 开发工具;
2. 掌握网络编程的基本理论。

16.4 程序代码

1. 使用 URL 类获取网络信息与资源。
(1)获取 URL 信息。
```
import java.net.*;
import java.io.*;
public class URLTest{
```

```java
public static void main(String[] args){
    URL url=null;
    InputStream is;
    try{
        url=new URL("http://localhost/index.html");
        is=url.openStream();
        int c;
        try{
            while((c=is.read())!=-1)
                System.out.print((char)c);
        }catch(IOException e){
        }finally{
            is.close();
        }
    }catch(MalformedURLException e){
        e.printStackTrace();
    }catch(IOException e){
        e.printStackTrace();
    }
    System.out.println("文件名:"+url.getFile());
    System.out.println("主机名:"+url.getHost());
    System.out.println("端口号:"+url.getPort());
    System.out.println("协议名:"+url.getProtocol());
}
}
```

程序运行结果如图 16.2 所示。

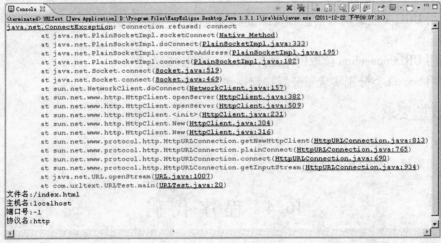

图 16.2 程序运行结果

（2）利用 URL 类获取网络资源。在本实验中，使用 URL 类来获取百度网站首页的内容并在控制台屏幕中显示出来。

import java.net.*;

第16章 网络程序设计

```
import java.io.*;
public class URL2{
    public static void main(String[] args) throws Exception{
        URL web=new URL("http://www.baidu.com/index.html");
        BufferedReader in=new BufferedReader(new
            InputStreamReader(web.openStream()));
        String inputLine;
        while((inputLine=in.readLine())!=null)System.out.println(inputLine);
        in.close();
    }
}
```

程序运行结果如图16.3所示。

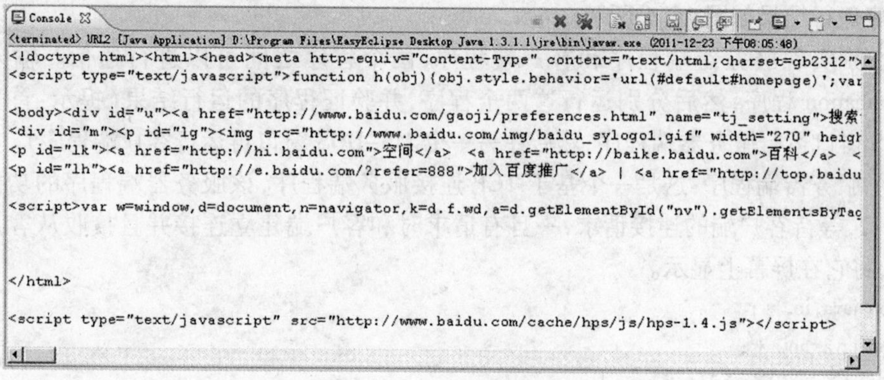

图16.3 程序运行结果

2. 利用URLConnection读取网络图片并保存到本地磁盘上。

```
import java.io.FileOutputStream;
import java.io.InputStream;
import java.io.OutputStream;
import java.net.URL;
import java.net.URLConnection;
public class URLDownloadPic{
    public static void main(String[] args) throws Exception{
        download("http://www.pestaola.gr/img/java_logo.png","logo.png");
    }
    public static void download(String urlString,String filename) throws Exception{
        //构造URL
        URL url=new URL(urlString);
        //打开连接
        URLConnection con=url.openConnection();
        //输入流
        InputStream is=con.getInputStream();
        //1K的数据缓冲
        byte[] bs=new byte[1024];
        //读取到的数据长度
```

```java
    int len;
    //输出的文件流
    OutputStream os = new FileOutputStream(filename);
    //开始读取
    while ((len = is.read(bs)) ! = -1) {
        os.write(bs, 0, len);
    }
    //完毕,关闭所有链接
    os.close();
    is.close();
}
```

图 16.4　程序运行结果

程序运行结果如图 16.4 所示。

3. 分别实现服务器端程序和客户端程序,两个程序之间可以进行通信。输入这两个 JavaApplication 程序,然后分别运行这两个程序,并验证程序的运行结果(提示:首先在一个 MS-DOS 窗口运行服务器端程序,然后在另一个 MS-DOS 窗口再运行客户端程序)。

(1) 服务器端程序:这是一个基于 TCP 连接服务端程序,该服务器端程序的功能是监听 8080 端口,等待客户端的连接请求,一旦有请求时和客户端建立连接并且接收从客户端发来的消息,将它在屏幕上显示。

```java
import java.io.*;
import java.net.*;
public class TCPSingleUserServer {
    //设置服务器的端口号,它应该大于 1024
    public static final int PORT = 8080;
    public static void main(String[] args) throws IOException {
        ServerSocket server = new ServerSocket(PORT);
        System.out.println("Started: " + server);
        try {
            // 等待接受用户连接
            Socket socket = server.accept();
            try {
                System.out.println("Connection accepted: " + socket);
                BufferedReader in = new BufferedReader(new
                    InputStreamReader(socket.getInputStream()));
                PrintWriter out = new PrintWriter(new BufferedWriter(new
                    OutputStreamWriter(socket.getOutputStream())), true);
                while (true) {
                    String str = in.readLine();
                    if (str.equals("END")) break;
                    System.out.println("Echoing: " + str);
                    out.println(str);
                }
            }
        }
```

```
            finally {
                System.out.println("closing...");
                socket.close();
            }
        }
        finally {
            server.close();
        }
    }
}
```

(2)客户端程序:客户端创建Socket对象连接到地址为127.0.0.1(本地主机的IP地址)的服务器端程序,端口为8080,并从键盘输入一行信息,发送到服务器,然后接受服务器的返回信息,最后结束会话。

```
import java.net.*;
import java.io.*;
public class SocketClient {
    Socket socket;
    BufferedReader in;
    PrintWriter out;
    public SocketClient() {
        try {
            socket = new Socket("127.0.0.1", 8080);
            in = new BufferedReader(new InputStreamReader(socket
                    .getInputStream()));
            out = new PrintWriter(socket.getOutputStream(), true);
            BufferedReader line = new BufferedReader(new InputStreamReader(
                    System.in));
            out.println(line.readLine());
            line.close();
            out.close();
            in.close();
            socket.close();
        } catch (IOException e) {
        }
    }
    public static void main(String[] args) {
        new SocketClient();
    }
}
```

程序运行结果如图16.5所示。

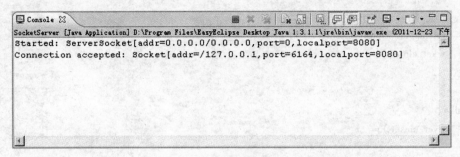

图 16.5 程序运行结果

第17章

综合案例

17.1 蜘蛛纸牌

一、总体设计

蜘蛛纸牌游戏由4个部分组成,分别是 Spider.java,SpiderMenuBar.java,PKCard.java 和 AboutDialog.java。

1. SpiderMenuBar.java

SpiderMenuBar.java 包含名为 SpiderMenuBar 的 public 类,其主要功能是生成蜘蛛纸牌游戏的菜单栏,实现菜单栏中各个组件的事件侦听。主要包括三个模块:图形用户界面的构建;组件监听接口的实现;显示可执行操作的线程。

2. PKCard.java

PKCard.java 包含名为 PKCard 的 public 类,其主要功能是定义纸牌的属性,包括名称、位置等相关信息,并通过相关方法实现了纸牌的移动等。

3. AboutDialog.java

AboutDialog.java 包含名为 AboutDialog 的 public 类,其主要功能是生成蜘蛛纸牌游戏的帮助栏。

4. Spider.java

Spider.java 包含名为 Spider 的 public 类,其主要功能是生成蜘蛛纸牌游戏的框架,实现游戏中的方法,包括纸牌的随机生成、位置的摆放等。

二、代码实现

1. SpiderMenuBar.java

```
import javax.swing.JMenuBar;
import javax.swing.JMenu;
import javax.swing.JMenuItem;
import javax.swing.JRadioButtonMenuItem;
import javax.swing.ButtonGroup;
public class SpiderMenuBar extends JMenuBar{
Spider main=null;
```

```java
JMenu jNewGame = new JMenu("游戏");
JMenu jHelp = new JMenu("帮助");
JMenuItem jItemAbout = new JMenuItem("关于");
JMenuItem jItemOpen = new JMenuItem("开局");
JMenuItem jItemPlayAgain = new JMenuItem("重新发牌");
JRadioButtonMenuItem jRMItemEasy = new JRadioButtonMenuItem("简单:单一花色");
JRadioButtonMenuItem jRMItemNormal = new JRadioButtonMenuItem("中级:双花色");
JRadioButtonMenuItem jRMItemHard = new JRadioButtonMenuItem("高级:四花色");
JMenuItem jItemExit = new JMenuItem("退出");
JMenuItem jItemValid = new JMenuItem("显示可行操作");
public SpiderMenuBar(Spider spider){
    this.main = spider;
    jNewGame.add(jItemOpen);
    jNewGame.add(jItemPlayAgain);
    jNewGame.add(jItemValid);
    jNewGame.addSeparator();
    jNewGame.add(jRMItemEasy);
    jNewGame.add(jRMItemNormal);
    jNewGame.add(jRMItemHard);
    jNewGame.addSeparator();
    jNewGame.add(jItemExit);
    ButtonGroup group = new ButtonGroup();
    group.add(jRMItemEasy);
    group.add(jRMItemNormal);
    group.add(jRMItemHard);
    jHelp.add(jItemAbout);
    this.add(jNewGame);
    this.add(jHelp);
    jItemOpen.addActionListener(new java.awt.event.ActionListener(){
        public void actionPerformed(java.awt.event.ActionEvent e){
            main.newGame();
        }
    });
    jItemPlayAgain.addActionListener(new java.awt.event.ActionListener(){
        public void actionPerformed(java.awt.event.ActionEvent e){
            if(main.getC()<60){
                main.deal();
            }
        }
    });
    jItemValid.addActionListener(new java.awt.event.ActionListener(){
        public void actionPerformed(java.awt.event.ActionEvent e){
            new Show().start();
        }
```

```java
});
jItemExit.addActionListener(new java.awt.event.ActionListener(){
    public void actionPerformed(java.awt.event.ActionEvent e){
        main.dispose();
        System.exit(0);
    }
});
jRMItemEasy.setSelected(true);
jRMItemEasy.addActionListener(new java.awt.event.ActionListener(){
    public void actionPerformed(java.awt.event.ActionEvent e){
        main.setGrade(Spider.EASY);
        main.initCards();
        main.newGame();
    }
});
jRMItemNormal.addActionListener(new java.awt.event.ActionListener(){
    public void actionPerformed(java.awt.event.ActionEvent e){
        main.setGrade(Spider.NATURAL);
        main.initCards();
        main.newGame();
    }
});
jRMItemNormal.addActionListener(new java.awt.event.ActionListener(){
    public void actionPerformed(java.awt.event.ActionEvent e){
        main.setGrade(Spider.HARD);
        main.initCards();
        main.newGame();
    }
});
jNewGame.addMenuListener(new javax.swing.event.MenuListener(){
    public void menuSelected(javax.swing.event.MenuEvent e){
        if(main.getC()<60){
            jItemPlayAgain.setEnabled(true);
        }
        else{
            jItemPlayAgain.setEnabled(false);
        }
    }
    public void menuDeselected(javax.swing.event.MenuEvent e){}
    public void menuCanceled(javax.swing.event.MenuEvent e){}
});
jItemAbout.addActionListener(new java.awt.event.ActionListener(){
    public void actionPerformed(java.awt.event.ActionEvent e){
        new AboutDialog();
```

```
      }
    });
  }
  class Show extends Thread{
    public void run(){
      main.showEnableOperator();
    }
  }
}
```

2. PKCard.java

```
import java.awt.*;
import java.awt.event.*;
import javax.swing.*;
public class PKCard extends JLabel implements MouseListener,MouseMotionListener{
  Point point=null;
  Point initPoint=null;
  int value=0;
  int type=0;
  String name=null;
  Container pane=null;
  Spider main=null;
  boolean canMove=false;
  boolean isFront=false;
  PkCard previousCard=null;
  public void mouseClicked(MouseEvent arg0){}
  public void flashCard(PKCard card){
    new Flash(card).start();
    if(main.getNextCard(card)!=null){
      card.flashCard(main.getNextCard(card));
    }
  }
  class Flash extends Thread{
    private PKCard card=null;
    public Flash(PKCard card){
      this.card=card;
    }
    public void run(){
      boolean is=false;
      ImageIcon icon=new ImageIcon("images/white.gif");   //使用图片
      for(int i=0;i<4;i++){
        try{
          Thread.sleep(200);
        }
        catch(InterruptedException e){
```

```java
            e.printStackTrace();
        }
        if(is){
            this.card.turnFront();
            is = ! is;
        }
        else
        {this.card.setIcon(icon);
         is! = is;}
        card.updateUI();
      }
   }
}
public void mousepressed(MouseEvent mp){
    point=mp.getPoint();
    main.setNA();
    this.previousCard=main.getPreviousCard(this);
}
public void mouseReleased(MouseEvent mr){
    Point point=((JLabel)mr.getSource()).getLocation();
    int n=this.whichColumnAvailable(point);
    if(n==-1||n==this.whichColumnAvailable(this.initPoint)){
        this.setNextCardLocation(null);
        main.table.remove(this.getLocation());
        this.setLocation(this.initPoint);
        main.table.put(this.initPoint,this);
        return;
    }
    point=main.getLastCardLocation(n);
    boolean isEmpty=false;
    PKCard card=null;
    if(point==null){
        point=main.getGroundLabelLocation(n);
        isEmpty=true;
    }
    else
    {
        card=(PKCard)main.tabel.get(point);
    }
    if(isEmpty||(this.value+1==card.getCardValue())){
        point.y+=40;
        if(isEmpty) point.y-=20;
        this.setNextCardLocation(point);
        main.table.remove(this.getLocation());
```

```java
          point.y-=20;
          this.setLocation(point);
          main.table.put(point,this);
          this.initPoint=point;
          if(this.previousCard! =null){
             this.previousCard.turnFront();
             this.previousCard.setCanMove(true);
          }
          this.setCanMove(true);
       }
       else{
          this.setNextCardLocation(null);
          main.table.emove(this.getLocation());
          this.setLocation(this.initPoint);
          main.table.put(this.initPoint,this)
       };
       return;
    }
    point=main.getLastCardLocation(n);
    card=(PKCard)main.table.get(point);
    if(card.getCardValue()= =1){
       point.y-=240;
       card=(PKCard)main.tabel.get(point);
       if(card! =null&&card.isCardCanMove()){
          main.haveFinish(n);
       }
    }
  }
}
public void setNextCardLocation(Point point){
    PKCard card=main.getNextCard(this);
    if(card! =null){
       if(point= =null){
          card.setNextCardLocation(null);
          main.table.remove(card.getLocation());
          card.setLocation(card.initPoint);
          main.table.put(card.initPoint,card);
            }
       else{
          point=new Point(point);
          point.y+=20;
          card.setNextCardLocation(point);
          point.y-=20;
          main.table.remove(card.getLocation());
          card.setLocation(point);
```

· 250 ·

```java
            main.table.put(card.getLocation(),card);
            card.initPoint=card.getLocation();
        }
    }
}
public int whichColumnAvailable(Point point){
    int x=point.x;
    int y=point.y;
    int a=(x-20)/101;
    int b=(x-20)%101;
    if(a!=9)
    {
        if(b>30&&b<=71){
            a=-1;
            }else if(b>71){
                a++;
            }
    }
    else if(b>71){
        a=-1;
    }
    if(a!=-1){
        Point p=main.getLastCardLocation(a);
        if(p==null)p=main.getgroundLabelLocation(a);
        b=y-p.y;
        if(b<=-96||b>=96){
            a=-1;
        }
    }
    return a;
}
public void mouseEntered(MouseEvent arg0){}
public void mouseExited(MouseEvent arg0){}
public void mouseDragged(MouseEvent arg0){
    if(canMove){
        int x=0;
        int y=0;
        Point p=arg0.getPoint();
        x=p.x-point.x;
        y=p.y-point.y;
        this.moving(x,y);
    }
}
public void moving(int x,int y){
```

```java
    PKCard card = main.getNextCard(this);
    Point p = this.getLocation();
    pane.setComponentZOrder(this, 1);
    main.table.remove(p);
    p.x += x;
    p.y += y;
    this.setLocation(p);
    main.table.put(p, this);
    if(card! = null) card.moving(x, y);

}
public void mouseMoved(MouseEvent arg0) {

}
public PKCard(String name, Spider spider) {
    super();
    this.type = new Integer(name.substring(0, 1)).intValue();
    this.value = new Integer(name.substring(2)).intValue();
    this.name = name;
    this.main = spider;
    this.pane = this.main.getContentPane();
    this.addMouseListener(this);
    this.addMouseMotionListener(this);
    this.setIcon(new ImageIcon("images/rear.gif"));
    this.setSize(71, 96);
    this.setVisible(true);
}
public void turnFront() {
    this.setIcon(new ImageIcon("images/" + name + ".gif"));
    this.isFront = true;
}
public void turnRear() {
    this.setIcon(new ImageIcon("images/rear.gif"));
    this.isFront = false;
    this.canMove = false;
}
public void moveto(Point point) {
    this.setLocation(point);
    this.initPoint = point;
}
public void setCanMove(boolean can) {
    this.canMove = can;
    PKCard card = main.getpreviousCard(this);
    if(card! = null && card.isCardFront()) {
```

```
            if(! can){
               if(! card.isCardcanMove()){
                  return;
               }
               else{
                  card.setCanMove(can);
               }
            }
            else{
               if(this.value+1==card.getCardValue()&&this.type==card.getCardType()){
                  card.setCanMove(can);
               }else
               {card.setCanMove(false);}
            }
         }
      }
   }
   public boolean isCardFront(){
      return this.isFront;
   }
   public boolean isCardCanMove(){
      return this.canMove;
   }
   public int getCardValue(){
      return value;
   }
   public int getCardType(){
      return type;
   }
```

3. AboutDialog.java

```
import javax.swing.*;
import java.awt.*;
public class AboutDialog extends JDialog{
   JPanel jMainPane=new JPanel();
   JTabbedPane jTabbedPane=new JTabbedPane();
   private JPanel jPanel1=new JPanel();
   private JPanel jPanel2=new JPanel();
   private JTextArea jt1=new JTextArea("将电脑多次分发给你的牌按照相同的花色由大到小排列起来。直到桌面上的牌全都消失。");
   private JTextArea jt2=new JTextArea("该游戏中,纸牌的图片来自于windowxp的纸牌游戏,图片权属于原作者所有!");
   public AboutDialog(){
      setTitle("蜘蛛牌");
      setSize(300,200);
```

```java
        setResizable(false);
        setDefaultCloseOperation(WindowConstants.DISPOSE_ON_CLOSE);
        Container c = this.getContentPane();
        jt1.setSize(260,200);
        jt2.setSize(260,200);
        jt1.setEditable(false);
        jt2.setEditable(false);
        jt1.setLineWrap(true);
        jt2.setLineWrap(true);
        jt1.setFont(new Font("楷体_GB2312",java.awt.Font.BOLD,13));
        jt2.setFont(new Font("楷体_GB2312",java.awt.Font.BOLD,13));
        jt2.setForeground(Color.black);
        jPanel1.add(jt1);
        jPanel2.add(jt2);
        jTabbedPane.setSize(300,200);
        jTabbedPane.addTab("游戏规则", null,jPanel1,null);
        jTabbedPane.addTab("声明", null,jPanel2,null);
        jMainPane.add(jTabbedPane);
        c.add(jMainPane);
        pack();
        this.setVisible(true);
    }
}
```

4. Spider.java

```java
import java.awt.*;
import java.awt.event.*;
import javax.swing.*;
import java.util.*;
public class Spider extends JFrame{
    public static final int EASY=1;
    public static final int NATURAL=2;
    public static final int HARD=3;
    private int grade=Spider.EASY;
    private Container pane=null;
    private PKCard cards[] = new PKCard[104];
    private JLabel clickLabel=null;
    private int c=0;
    private int n=0;
    private int a=0;
    private int finish=0;
    Hashtable table=null;
    private JLabel groundLabel[] =null;
    public static void main(String[] args){
        Spider spider=new Spider();
```

```
        spider.setVisible(true);
    }
    public Spider(){
        setTitle("蜘蛛牌");
        setDefaultCloseOperation(javax.swing.JFrame.EXIT_ON_CLOSE);
        setSize(1024,742);
        setJMenuBar(new SpiderMenuBar(this));
        pane=this.getContentPane();
        pane.setBackground(new Color(0,112,26));
        pane.setLayout(null);
        clickLabel=new JLabel();
        clickLabel.setBounds(883,606,121,96);
        pane.add(clickLabel);
        clickLabel.addMouseListener(new MouseAdapter(){
            public void mouseReleased(MouseEvent me){
                if(c<60){
                    Spider.this.deal();
                }
            }
        });
        this.initCards();
        this.randomCards();
        this.setCardsLocation();
        groundLabel=new JLabel[10];
        int x=20;
        for(int i=0;i<10;i++){
            groundLabel[i]=new JLabel();
     groundLabel[i].setBorder(javax.swing.BorderFactory.createEtchedBorder(javax.swing.border.
EtchedBorder.RAISED));
            groundLabel[i].setBounds(x,25,71,96);
            x+=101;
            this.pane.add(groundLabel[i]);
        }
        this.setVisible(true);
        this.deal();
        this.addKeyListener(new KeyAdapter(){
            class Show extends Thread{
                public void run(){
                    Spider.this.showEnabledOperator();
                }
            }
            public void keyPressed(KeyEvent e){
                if(finish!=8)if(e.getKeyCode()==KeyEvent.VK_D&&c<60){
                    spider.this.deal();
```

```java
      }
      else if(e.getKeyCode()==KeyEvent.VK_M){
        new Show().start();
      }
    }
  });
}
public void newGame(){
  this.randomCards();
  this.setCardsLocation();
  this.setGroundLabelZorder();
  this.deal();
}
public int getC(){
  return c;
}
public void setGrade(int grade){
  this.grade=grade;
}
public void initCards(){
  if(cards[0]!=null){
    for(int i=0;i<104;i++)
      pane.remove(cards[i]);
  }
  int n=0;
  if(this.grade==Spider.EASY){
    n=1;
  }
  else if(this.grade==Spider.NATURAL){
    n=2;
  }
  else
  {n=4;}
  for(int i=1;i<=8;i++){
    for(int j=1;j<=13;j++)
      cards[(i-1)*13+j-1]=new PKCard((i%n+1)+"-"+j,this);
  }
  this.randomCards();
}
public void randomCards(){
  PKCard temp=null;
  for(int i=0;i<52;i++){
```

```java
      int a=(int)(Math.random()*104);
      int b=(int)(Math.random()*104);
      temp=cards[a];
      cards[a]=cards[b];
      cards[b]=temp;
   }
}
public void setNA(){
   a=0;
   n=0;
}
public void setCardsLocation(){
   table=new Hashtable();
   c=0;
   finish=0;
   n=0;
   a=0;
   int x=883;
   int y=580;
   for(int i=0;i<6;i++){
      for(int j=0;j<10;j++){
         int n=i*10+j;
         pane.add(cards[n]);
         cards[n].turnRear();
         cards[n].moveto(new Point(x,y));
         table.put(new Point(x,y),cards[n]);
      }
      x+=10;
   }
   x=20;
   y=45;
   for(int i=10;i>5;i--){
      for(int j=0;j<10;j++){
         int n=i*10+j;
         if(n>=104)continue;
         pane.add(cards[n]);
         cards[n].turnRear();
         cards[n].moveto(new Point(x,y));
         table.put(new Point(x,y),cards[n]);
         x+=101;
      }
      x=20;
      y-=5;
   }
```

```java
    }
    public void showEnableOperator() {
        int x=0;
        out: while(true) {
            Point point=null;
            PKCard card=null;
            do{
                if(point! =null) {
                    n++;
                }
                point=this.getLastCardLocation(n);
                while(point==null) {
                    point=this.getLastCardLocation(++n);
                    if(n==10)n=0;
                    x++;
                    if(x==10)break out;
                }
                card=(PKCard)this.table.get(point);
            }
            while(! card.isCardCanMove());
while(this.getpreviousCard(card)! =null&&this.getPreviousCard(card).isCardCanMove()) {
                card=this.getPreviousCard(card);
            }
            if(a==10) {
                a=0;
            }
            for(;a<10;a++) {
                if(a! =n) {
                    Point p=null;
                    PKCard c=null;
                    do{
                        if(p! =null) {a++;}
                        p=this.getLastcardLocation(a);
                        int z=0;
                        while(p==null) {

                        }
                    }
                }
            }
        }
    }
}
```

```java
/*
 * *返回值:void *方法:游戏运行
 */
public void deal()
{
    this.setNA();
    //判断10列中是否空列
    for (int i=0; i < 10; i++){
        if (this.getLastCardLocation(i)==null){
            JOptionPane.showMessageDialog(this,"有空位不能发牌!","提示",
                JOptionPane.WARNING_MESSAGE);
            return;
        }
    }
    int x=20;

for (int i=0; i < 10; i++){
    Point lastPoint=this.getLastCardLocation(i);
    //这张牌应"背面向上"
if (c==0){
        lastPoint.y+=5;
}
    //这张牌应"正面向上"
else{
        lastPoint.y+=20;
}

table.remove(cards[c+i].getLocation());
        cards[c+i].moveto(lastPoint);
        table.put(new Point(lastPoint), cards[c+i]);
        cards[c+i].turnFront();
        cards[c+i].setCanMove(true);

// 将组件card移动到容器中指定的顺序索引。
this.pane.setComponentZOrder(cards[c+i], 1);

Point point=new Point(lastPoint);
        if (cards[c+i].getCardValue()==1){
            int n=cards[c+i].whichColumnAvailable(point);
            point.y -=240;
            PKCard card=(PKCard) this.table.get(point);
            if (card != null && card.isCardCanMove()){
                this.haveFinish(n);
            }
```

```
            }
            x+=101;
        }
        c+=10;
    }

    /*
     * *返回值:PKCard 对象  *方法:获得 card 上面的那张牌
     */
    public PKCard getPreviousCard(PKCard card){
        Point point=new Point(card.getLocation());
        point.y -=5;
        card=(PKCard) table.get(point);
        if(card !=null){
            return card;
        }
        point.y -=15;
        card=(PKCard) table.get(point);
        return card;
    }

    /**
     * *返回值:PKCard 对象  *方法:取得 card 下面的一张牌
     */
    public PKCard getNextCard(PKCard card){
        Point point=new Point(card.getLocation());
        point.y+=5;
        card=(PKCard) table.get(point);
        if(card !=null)
            return card;
        point.y+=15;
        card=(PKCard) table.get(point);
        return card;
    }

    /**
     * *返回值:Point 对象  *方法:取得第 column 列最后一张牌的位置
     */
    public Point getLastCardLocation(int column){
        Point point=new Point(20+column * 101, 25);
        PKCard card=(PKCard) this.table.get(point);
        if(card==null) return null;
        while(card !=null){
            point=card.getLocation();
```

```java
            card = this.getNextCard(card);
        }
        return point;
    }

    public Point getGroundLabelLocation(int column){
        return new Point(groundLabel[column].getLocation());
    }

    /*
     * *返回值:void  *方法:放置 groundLable 组件
     */
    public void setGroundLabelZOrder(){
        for(int i=0; i < 10; i++){
            //将组件 groundLable 移动到容器中指定的顺序索引。顺序(105+i)确定了绘制组件的顺序;具有最
            //高顺序的组件将第一个绘制,而具有最低顺序的组件将最后一个绘制。在组件重叠的地方,具有较低顺序的
            //组件将覆盖具有较高顺序的组件。
            pane.setComponentZOrder(groundLabel[i], 105+i);
        }
    }


    /*
     * *返回值:void  *方法:判断纸牌的摆放是否完成
     */
    public void haveFinish(int column){
        Point point = this.getLastCardLocation(column);
        PKCard card = (PKCard) this.table.get(point);
        do{
            this.table.remove(point);
            card.moveto(new Point(20+finish * 10, 580));
            //将组件移动到容器中指定的顺序索引
            pane.setComponentZOrder(card, 1);
            //将纸牌新的相关信息存入 Hashtable
            this.table.put(card.getLocation(), card);
            card.setCanMove(false);
            point = this.getLastCardLocation(column);
            if(point == null)
                card = null;
            else
                card = (PKCard) this.table.get(point);
        }
        while(card != null && card.isCardCanMove());
        finish++;
        //如果8付牌全部组合成功,则显示成功的对话框
```

```
        if (finish = = 8) {
            JOptionPane. showMessageDialog( this, "恭喜你,顺利通过!", "成功",
                    JOptionPane. PLAIN _ MESSAGE );
        }
        if ( card！= null) {
            card. turnFront( );
            card. setCanMove(true );
        }
    }
}
```

三、程序的运行与发布

1. 运行程序

将文件 Spider. java、SpiderMenuBar. java、PKCard. java、AboutDialog. java 及所需要的 images 图像文件夹保存到同一个文件夹中。利用 javac 命令对文件进行编辑,使用的命令如下:

javac Spider. java

之后,利用 Java 命令执行程序,使用的命令如下:

javaSpider

2. 发布程序

要发布此应用程序,需要将应用程序打包。使用 jar. exe,可以把应用程序中涉及和图片压缩成一个 jar 文件,这样便可以发布程序。

首先编写一个清单文件,名为 MANIFEST. MF,其代码如下:

Manifest-Version：1.0

Created-By：1.6.0(Sun microsystem Inc.)

Main-class：Spider

将此清单文件保存到 C:\Javawork\ch05 文件夹中。

然后,使用如下命令生成 jar 文件:

jar cfm Spder. jar MANIFEST. MF ＊. class

其中参数 c 表示要生成一个新的 jar 文件;f 表示要生成的 jar 文件的名字;m 表示清单文件的名字。

如果机器安装这 WinRAR 解压软件,并将. jar 文件与解压缩软件作了关联,那么 Spider. jar 文件的类型是 WinRAR,使得 Java 程序无法运行。因此,我们在发布软件时,还应该再写一个有如下内容的 bat 文件(Spider. bat):

javaw-jar Spider. jar

然后可以通过双击 Spider. bat 来运行程序。

17.2　Java 聊天室

一、功能描述

本案例中,我们利用 Java 实现基于 C/S 模式的聊天室程序。聊天室共分服务器端和客户端两部分,服务器端程序主要负责侦听客户端发来的消息,客户端需登录到服务端才可以实现正常的聊天功能。

1. 服务器端的主要功能

(1)在特定端口上进行侦听,等待客户端连接。

(2)用户可以配置服务端的侦听端口,默认端口为 8888。

(3)向已经连接到服务器端的用户发送系统消息。

(4)统计在线人数。

(5)当停止服务时,断开所有的用户连接。

2. 客户端的主要功能

(1)连接到已经开启聊天服务的服务端。

(2)用户可以配置要连接服务器端的 IP 地址与端口号。

(3)用户可以配置连接后显示的用户名。

(4)当服务器端开启时,用户可以随时登录与注销。

(5)用户可以向所有人或者某一个人发送消息。

二、总体设计

1. 聊天室服务器端

聊天室服务器端主要包括 7 个文件,它们的功能如下:

(1)ChatServer.java。

该文件包含名为了 ChatServer 的 public 类,其主要功能是定义服务器端的界面,添加事件侦听与事件处理。ChatServe 类调用 ServerListen 类来实现服务端用户上线与下线的侦听,调用 ServerReceive 类来实现服务器端的消息收发。

(2)ServerListen.java。

该类实现服务端用户上线与下线的侦听。该类对用户上线和下线的侦听是通过调用用户链表类(UserLinkList)来实现的,当用户的上线与下线情况发生变化时,该类会对主类的界面进行相应的修改。

(3)ServerReceive.java。

该类是实现服务器消息收发的类。该类分别定义了向某用户及所有人发送消息的方法。发送的消息会显示在主界面类的界面上。

(4)PortConf.java。

该类继承自 JDialog,是用户对服务器端侦听端口进行修改配置的类。

(5)Node.java。

用户链表的节点类,定义了链表中的用户。该类与前面所讲的链表节点 Node 类的功能相当。

(6)UserLinkList.java。

用户链表节点的具体实现类。该类通过构造函数构造用户链表,定义了添加用户、删除用户、返回用户数、根据用户名查找用户和根据索引查找用户这 5 个方法。

(7)Help.java。

服务端程序的帮助类。

2. 聊天室客户端

聊天室客户端主要包括 5 个文件,它们的功能如下:

(1) ChatClient.java。

该文件包含名为 ChatClient 的 public 类,其主要功能是定义客户端的界面,添加事件侦听与事件处理。该类定义了 Connect( ) 与 DisConnect( ) 方法实现与服务器的连接和断开连接。当用户登录到指定的服务器时,ChatClient 类调用 ClientReceive 类实现消息收发,同时 ChatClient 类还定义了 SendMessage( ) 方法来向其他用户发送带有表情的消息或者悄悄话。

(2) ClientReceive.java。

该类是实现服务器端与客户端消息收发的类。

(3) ConnectConf.java。

该类继承自 JDialog,是用户对所要连接的服务器 IP 及侦听端口进行修改配置的类。

(4) UserConf.java。

该类继承自 JDialog,是用户对连接到服务器时所显示的用户名进行修改配置的类。

(5) Help.java。

客户端程序的帮助类。

三、代码实现

服务端代码:

1. chatserver.java

```java
import java.awt.*;
import java.awt.event.*;
import javax.swing.*;
import javax.swing.event.*;
import java.net.*;
import java.io.*;
/*
*聊天服务端的主框架类
*/
public class ChatServer extends JFrame implements ActionListener{
public static int port=8888;//服务端的侦听端口
ServerSocket serverSocket;//服务端 Socket
Image icon;//程序图标

JComboBox combobox;//选择发送消息的接受者
JTextArea messageShow;//服务端的信息显示
JScrollPane messageScrollPane;//信息显示的滚动条
JTextField showStatus;//显示用户连接状态
JLabel sendToLabel,messageLabel;
JTextField sysMessage;//服务端消息的发送
JButton sysMessageButton;//服务端消息的发送按钮
UserLinkList userLinkList;//用户链表
//建立菜单栏
```

```java
JMenuBar jMenuBar=new JMenuBar();
//建立菜单组
JMenu serviceMenu=new JMenu("服务(V)");
//建立菜单项
JMenuItem portItem=new JMenuItem("端口设置(P)");
JMenuItem startItem=new JMenuItem("启动服务(S)");
JMenuItem stopItem=new JMenuItem("停止服务(T)");
JMenuItem exitItem=new JMenuItem("退出(X)");
JMenu helpMenu=new JMenu("帮助(H)");
JMenuItem helpItem=new JMenuItem("帮助(H)");
//建立工具栏
JToolBar toolBar=new JToolBar();
//建立工具栏中的按钮组件
JButton portSet;//启动服务端侦听
JButton startServer;//启动服务端侦听
JButton stopServer;//关闭服务端侦听
JButton exitButton;//退出按钮
//框架的大小
Dimension faceSize=new Dimension(400,600);
ServerListen listenThread;
JPanel downPanel;
GridBagLayout girdBag;
GridBagConstraints girdBagCon;
/**
 *服务端构造函数
 */
public ChatServer() {
    init();//初始化程序
    //添加框架的关闭事件处理
    this.setDefaultCloseOperation(JFrame.EXIT_ON_CLOSE);
    this.pack();
    //设置框架的大小
    this.setSize(faceSize);
    //设置运行时窗口的位置
    Dimension screenSize=Toolkit.getDefaultToolkit().getScreenSize();
    this.setLocation((int)(screenSize.width-faceSize.getWidth())/2,
        (int)(screenSize.height-faceSize.getHeight())/2);
    this.setResizable(false);
    this.setTitle("聊天室服务端");//设置标题
    //程序图标
    icon=getImage("icon.gif");
    this.setIconImage(icon);//设置程序图标
    show();
    //为服务菜单栏设置热键'V'
```

```java
serviceMenu.setMnemonic('V');
//为端口设置快捷键为 ctrl+p
portItem.setMnemonic('P');
portItem.setAccelerator(KeyStroke.getKeyStroke(KeyEvent.VK_P,InputEvent.CTRL_MASK));
//为启动服务快捷键为 ctrl+s
startItem.setMnemonic('S');
startItem.setAccelerator(KeyStroke.getKeyStroke(KeyEvent.VK_S,InputEvent.CTRL_MASK));
//为端口设置快捷键为 ctrl+t
stopItem.setMnemonic('T');
stopItem.setAccelerator(KeyStroke.getKeyStroke(KeyEvent.VK_T,InputEvent.CTRL_MASK));
//为退出设置快捷键为 ctrl+x
exitItem.setMnemonic('X');
exitItem.setAccelerator(KeyStroke.getKeyStroke(KeyEvent.VK_X,InputEvent.CTRL_MASK));
//为帮助菜单栏设置热键'H'
helpMenu.setMnemonic('H');
//为帮助设置快捷键为 ctrl+p
helpItem.setMnemonic('H');
helpItem.setAccelerator(KeyStroke.getKeyStroke(KeyEvent.VK_H,InputEvent.CTRL_MASK));
}
/**
 *程序初始化函数
 */
public void init() {
    Container contentPane = getContentPane();
    contentPane.setLayout(new BorderLayout());
    //添加菜单栏
    serviceMenu.add(portItem);
    serviceMenu.add(startItem);
    serviceMenu.add(stopItem);
    serviceMenu.add(exitItem);
    jMenuBar.add(serviceMenu);
    helpMenu.add(helpItem);
    jMenuBar.add(helpMenu);
    setJMenuBar(jMenuBar);
    //初始化按钮
    portSet = new JButton("端口设置");
    startServer = new JButton("启动服务");
    stopServer = new JButton("停止服务");
    exitButton = new JButton("退出");
    //将按钮添加到工具栏
    toolBar.add(portSet);
    toolBar.addSeparator();//添加分隔栏
    toolBar.add(startServer);
    toolBar.add(stopServer);
```

```
toolBar. addSeparator( );//添加分隔栏
toolBar. add( exitButton) ;
contentPane. add( toolBar, BorderLayout. NORTH) ;
//初始时,令停止服务按钮不可用
stopServer. setEnabled( false) ;
stopItem. setEnabled( false) ;
//为菜单栏添加事件监听
portItem. addActionListener( this) ;
startItem. addActionListener( this) ;
stopItem. addActionListener( this) ;
exitItem. addActionListener( this) ;
helpItem. addActionListener( this) ;
//添加按钮的事件侦听
portSet. addActionListener( this) ;
startServer. addActionListener( this) ;
stopServer. addActionListener( this) ;
exitButton. addActionListener( this) ;
combobox = new JComboBox( ) ;
combobox. insertItemAt("所有人", 0) ;
combobox. setSelectedIndex(0) ;
messageShow = new JTextArea( ) ;
messageShow. setEditable( false) ;
//添加滚动条
messageScrollPane = new JScrollPane( messageShow,
JScrollPane. VERTICAL _ SCROLLBAR _ AS _ NEEDED,
JScrollPane. HORIZONTAL _ SCROLLBAR _ AS _ NEEDED) ;
messageScrollPane. setPreferredSize( new Dimension(400, 400) ) ;
messageScrollPane. revalidate( ) ;
showStatus = new JTextField(35) ;
showStatus. setEditable( false) ;
sysMessage = new JTextField(24) ;
sysMessage. setEnabled( false) ;
sysMessageButton = new JButton( ) ;
sysMessageButton. setText("发送") ;
//添加系统消息的事件侦听
sysMessage. addActionListener( this) ;
sysMessageButton. addActionListener( this) ;
sendToLabel = new JLabel("发送至:") ;
messageLabel = new JLabel("发送消息:") ;
downPanel = new JPanel( ) ;
girdBag = new GridBagLayout( ) ;
downPanel. setLayout( girdBag) ;
girdBagCon = new GridBagConstraints( ) ;
girdBagCon. gridx = 0 ;
```

```
girdBagCon.gridy=0;
girdBagCon.gridwidth=3;
girdBagCon.gridheight=2;
girdBagCon.ipadx=5;
girdBagCon.ipady=5;
JLabel none=new JLabel("    ");
girdBag.setConstraints(none, girdBagCon);
downPanel.add(none);
girdBagCon=new GridBagConstraints();
girdBagCon.gridx=0;
girdBagCon.gridy=2;
girdBagCon.insets=new Insets(1, 0, 0, 0);
girdBagCon.ipadx=5;
girdBagCon.ipady=5;
girdBag.setConstraints(sendToLabel, girdBagCon);
downPanel.add(sendToLabel);
girdBagCon=new GridBagConstraints();
girdBagCon.gridx=1;
girdBagCon.gridy=2;
girdBagCon.anchor=GridBagConstraints.LINE_START;
girdBag.setConstraints(combobox, girdBagCon);
downPanel.add(combobox);
girdBagCon=new GridBagConstraints();
girdBagCon.gridx=0;
girdBagCon.gridy=3;
girdBag.setConstraints(messageLabel, girdBagCon);
downPanel.add(messageLabel);
girdBagCon=new GridBagConstraints();
girdBagCon.gridx=1;
girdBagCon.gridy=3;
girdBag.setConstraints(sysMessage, girdBagCon);
downPanel.add(sysMessage);
girdBagCon=new GridBagConstraints();
girdBagCon.gridx=2;
girdBagCon.gridy=3;
girdBag.setConstraints(sysMessageButton, girdBagCon);
downPanel.add(sysMessageButton);
girdBagCon=new GridBagConstraints();
girdBagCon.gridx=0;
girdBagCon.gridy=4;
girdBagCon.gridwidth=3;
girdBag.setConstraints(showStatus, girdBagCon);
downPanel.add(showStatus);
contentPane.add(messageScrollPane, BorderLayout.CENTER);
```

```java
contentPane.add(downPanel, BorderLayout.SOUTH);
//关闭程序时的操作
this.addWindowListener(new WindowAdapter() {
    public void windowClosing(WindowEvent e) {
        stopService();
        System.exit(0);
    }
});
}
/**
 *事件处理
 */
public void actionPerformed(ActionEvent e) {
    Object obj = e.getSource();
    if (obj == startServer || obj == startItem) { //启动服务端
        startService();
    } else if (obj == stopServer || obj == stopItem) { //停止服务端
        int j = JOptionPane.showConfirmDialog(this, "真的停止服务吗?", "停止服务",
            JOptionPane.YES_OPTION, JOptionPane.QUESTION_MESSAGE);
        if (j == JOptionPane.YES_OPTION) {
            stopService();
        }
    } else if (obj == portSet || obj == portItem) { //端口设置
        //调出端口设置的对话框
        PortConf portConf = new PortConf(this);
        portConf.show();
    } else if (obj == exitButton || obj == exitItem) { //退出程序
        int j = JOptionPane.showConfirmDialog(this, "真的要退出吗?", "退出",
            JOptionPane.YES_OPTION, JOptionPane.QUESTION_MESSAGE);
        if (j == JOptionPane.YES_OPTION) {
            stopService();
            System.exit(0);
        }
    } else if (obj == helpItem) { //菜单栏中的帮助
        //调出帮助对话框
        Help helpDialog = new Help(this);
        helpDialog.show();
    } else if (obj == sysMessage || obj == sysMessageButton) { //发送系统消息
        sendSystemMessage();
    }
}
/**
 *启动服务端
 */
```

```java
public void startService() {
    try {
        serverSocket=new ServerSocket(port, 10);
        messageShow.append("服务端已经启动,在"+port+"端口侦听...\n");
        startServer.setEnabled(false);
        startItem.setEnabled(false);
        portSet.setEnabled(false);
        portItem.setEnabled(false);
        stopServer.setEnabled(true);
        stopItem.setEnabled(true);
        sysMessage.setEnabled(true);
    } catch (Exception e) {
        //System.out.println(e);
    }
    userLinkList=new UserLinkList();
    listenThread=new ServerListen(serverSocket, combobox, messageShow,
        showStatus, userLinkList);
    listenThread.start();
}
/**
 * 关闭服务端
 */
public void stopService() {
    try {
        //向所有人发送服务器关闭的消息
        sendStopToAll();
        listenThread.isStop=true;
        serverSocket.close();
        int count=userLinkList.getCount();
        int i=0;
        while (i < count) {
            Node node=userLinkList.findUser(i);
            node.input.close();
            node.output.close();
            node.socket.close();
            i++;
        }
        stopServer.setEnabled(false);
        stopItem.setEnabled(false);
        startServer.setEnabled(true);
        startItem.setEnabled(true);
        portSet.setEnabled(true);
        portItem.setEnabled(true);
        sysMessage.setEnabled(false);
```

```java
messageShow.append("服务端已经关闭\n");
combobox.removeAllItems();
combobox.addItem("所有人");
} catch (Exception e) {
//System.out.println(e);
}
}
/**
 * 向所有人发送服务器关闭的消息
 */
public void sendStopToAll() {
    int count = userLinkList.getCount();
    int i = 0;
    while (i < count) {
        Node node = userLinkList.findUser(i);
        if (node == null) {
            i++;
            continue;
        }
        try {
            node.output.writeObject("服务关闭");
            node.output.flush();
        } catch (Exception e) {
            //System.out.println("$ $ $"+e);
        }
        i++;
    }
}
/**
 * 向所有人发送消息
 */
public void sendMsgToAll(String msg) {
    int count = userLinkList.getCount();//用户总数
    int i = 0;
    while (i < count) {
        Node node = userLinkList.findUser(i);
        if (node == null) {
            i++;
            continue;
        }
        try {
            node.output.writeObject("系统信息");
            node.output.flush();
            node.output.writeObject(msg);
```

```java
node.output.flush();
} catch (Exception e) {
//System.out.println("@@@"+e);
}
i++;
}
sysMessage.setText("");
}
/**
 * 向客户端用户发送消息
 */
public void sendSystemMessage() {
    String toSomebody=combobox.getSelectedItem().toString();
    String message=sysMessage.getText()+"\n";
    messageShow.append(message);
    //向所有人发送消息
    if(toSomebody.equalsIgnoreCase("所有人")) {
        sendMsgToAll(message);
    } else {
        //向某个用户发送消息
        Node node=userLinkList.findUser(toSomebody);
        try {
            node.output.writeObject("系统信息");
            node.output.flush();
            node.output.writeObject(message);
            node.output.flush();
        } catch (Exception e) {
            //System.out.println("!!!"+e);
        }
        sysMessage.setText("");//将发送消息栏的消息清空
    }
}
/**
 * 通过给定的文件名获得图像
 */
Image getImage(String filename) {
    URLClassLoader urlLoader=(URLClassLoader) this.getClass()
        .getClassLoader();
    URL url=null;
    Image image=null;
    url=urlLoader.findResource(filename);
    image=Toolkit.getDefaultToolkit().getImage(url);
    MediaTracker mediatracker=new MediaTracker(this);
    try {
```

```java
        mediatracker.addImage(image, 0);
        mediatracker.waitForID(0);
    } catch (InterruptedException _ex) {
        image = null;
    }
    if (mediatracker.isErrorID(0)) {
        image = null;
    }
    return image;
}

public static void main(String[] args) {
    ChatServer app = new ChatServer();
}
}
```

2. serverlisten.java

```java
import java.awt.*;
import java.awt.event.*;
import javax.swing.*;
import javax.swing.event.*;
import java.io.*;
import java.net.*;
/*
 *服务端的侦听类
 */
public class ServerListen extends Thread {
    ServerSocket server;
    JComboBox combobox;
    JTextArea textarea;
    JTextField textfield;
    UserLinkList userLinkList;//用户链表
    Node client;
    ServerReceive recvThread;
    public boolean isStop;
    /*
     *聊天服务端的用户上线于下线侦听类
     */
    public ServerListen(ServerSocket server, JComboBox combobox,
            JTextArea textarea, JTextField textfield, UserLinkList userLinkList) {
        this.server = server;
        this.combobox = combobox;
        this.textarea = textarea;
        this.textfield = textfield;
        this.userLinkList = userLinkList;
```

```java
isStop = false;
}
public void run() {
    while (! isStop && ! server.isClosed()) {
        try {
            client = new Node();
            client.socket = server.accept();
            client.output = new ObjectOutputStream(client.socket.getOutputStream());
            client.output.flush();
            client.input = new ObjectInputStream(client.socket.getInputStream());
            client.username = (String) client.input.readObject();
            // 显示提示信息
            combobox.addItem(client.username);
            userLinkList.addUser(client);
            textarea.append("用户 "+client.username+" 上线"+"\n");
            textfield.setText("在线用户"+userLinkList.getCount()+"人\n");
            recvThread = new ServerReceive(textarea, textfield, combobox, client, userLinkList);
            recvThread.start();
        } catch (Exception e) {
        }
    }
}
}
```

3. serverreceive.java

```java
import javax.swing.*;
import java.io.*;
import java.net.*;
/*
 * 服务器收发消息的类
 */
public class ServerReceive extends Thread {
    JTextArea textarea;
    JTextField textfield;
    JComboBox combobox;
    Node client;
    UserLinkList userLinkList;//用户链表
    public boolean isStop;
    public ServerReceive(JTextArea textarea, JTextField textfield,
        JComboBox combobox, Node client, UserLinkList userLinkList) {
        this.textarea = textarea;
        this.textfield = textfield;
        this.client = client;
        this.userLinkList = userLinkList;
        this.combobox = combobox;
```

```java
isStop = false;
}
public void run() {
//向所有人发送用户的列表
sendUserList();
while (! isStop && ! client.socket.isClosed()) {
try {
String type = (String) client.input.readObject();
if (type.equalsIgnoreCase("聊天信息")) {
String toSomebody = (String) client.input.readObject();
String status = (String) client.input.readObject();
String action = (String) client.input.readObject();
String message = (String) client.input.readObject();
String msg = client.username+" "+action+"对 "
+ toSomebody+" 说 : "+message+"\n";
if (status.equalsIgnoreCase("悄悄话")) {
msg = "[悄悄话] "+msg;
}
textarea.append(msg);
if (toSomebody.equalsIgnoreCase("所有人")) {
sendToAll(msg);//向所有人发送消息
} else {
try {
client.output.writeObject("聊天信息");
client.output.flush();
client.output.writeObject(msg);
client.output.flush();
} catch (Exception e) {
//System.out.println("###"+e);
}
Node node = userLinkList.findUser(toSomebody);
if (node != null) {
node.output.writeObject("聊天信息");
node.output.flush();
node.output.writeObject(msg);
node.output.flush();
}
}
} else if (type.equalsIgnoreCase("用户下线")) {
Node node = userLinkList.findUser(client.username);
userLinkList.delUser(node);
String msg = "用户 "+client.username+" 下线\n";
int count = userLinkList.getCount();
combobox.removeAllItems();
```

```
combobox.addItem("所有人");
int i=0;
while (i < count) {
node=userLinkList.findUser(i);
if (node==null) {
i++;
continue;
}
combobox.addItem(node.username);
i++;
}
combobox.setSelectedIndex(0);
textarea.append(msg);
textfield.setText("在线用户"+userLinkList.getCount()+"人\n");
sendToAll(msg);//向所有人发送消息
sendUserList();//重新发送用户列表,刷新
break;
}
} catch (Exception e) {
//System.out.println(e);
}
}
}
/*
*向所有人发送消息
*/
public void sendToAll(String msg) {
int count=userLinkList.getCount();
int i=0;
while (i < count) {
Node node=userLinkList.findUser(i);
if (node==null) {
i++;
continue;
}
try {
node.output.writeObject("聊天信息");
node.output.flush();
node.output.writeObject(msg);
node.output.flush();
} catch (Exception e) {
//System.out.println(e);
}
i++;
```

```
    }
  }
  /*
   *向所有人发送用户的列表
   */
  public void sendUserList() {
    String userlist = "";
    int count = userLinkList.getCount();
    int i = 0;
    while (i < count) {
      Node node = userLinkList.findUser(i);
      if (node = = null) {
        i++;
        continue;
      }
      userlist+ = node.username;
      userlist+ = '\n';
      i++;
    }
    i = 0;
    while (i < count) {
      Node node = userLinkList.findUser(i);
      if (node = = null) {
        i++;
        continue;
      }
      try {
        node.output.writeObject("用户列表");
        node.output.flush();
        node.output.writeObject(userlist);
        node.output.flush();
      } catch (Exception e) {
        //System.out.println(e);
      }
      i++;
    }
  }
}
```

4. proconf.java

```
import java.awt.*;
import javax.swing.border.*;
import java.net.*;
import javax.swing.*;
import java.awt.event.*;
```

```java
/**
 *生成端口设置对话框的类
 */
public class PortConf extends JDialog {
    JPanel panelPort = new JPanel();
    JButton save = new JButton();
    JButton cancel = new JButton();
    public static JLabel DLGINFO = new JLabel("默认端口号为:8888");
    JPanel panelSave = new JPanel();
    JLabel message = new JLabel();
    public static JTextField portNumber;
    public PortConf(JFrame frame) {
        super(frame, true);
        try {
            jbInit();
        } catch (Exception e) {
            e.printStackTrace();
        }
        //设置运行位置,使对话框居中
        Dimension screenSize = Toolkit.getDefaultToolkit().getScreenSize();
        this.setLocation((int)(screenSize.width - 400) / 2+50,
            (int)(screenSize.height - 600) / 2+150);
        this.setResizable(false);
    }
    private void jbInit() throws Exception {
        this.setSize(new Dimension(300, 120));
        this.setTitle("端口设置");
        message.setText("请输入侦听的端口号:");
        portNumber = new JTextField(10);
        portNumber.setText(""+ChatServer.port);
        save.setText("保存");
        cancel.setText("取消");
        panelPort.setLayout(new FlowLayout());
        panelPort.add(message);
        panelPort.add(portNumber);
        panelSave.add(new Label("                "));
        panelSave.add(save);
        panelSave.add(cancel);
        panelSave.add(new Label("                "));
        Container contentPane = getContentPane();
        contentPane.setLayout(new BorderLayout());
        contentPane.add(panelPort, BorderLayout.NORTH);
        contentPane.add(DLGINFO, BorderLayout.CENTER);
        contentPane.add(panelSave, BorderLayout.SOUTH);
```

```java
//保存按钮的事件处理
save.addActionListener(new ActionListener() {
public void actionPerformed(ActionEvent a) {
int savePort;
try {
savePort = Integer.parseInt(PortConf.portNumber.getText());
if (savePort < 1 || savePort > 65535) {
PortConf.DLGINFO.setText("侦听端口必须是0-65535之间的整数!");
PortConf.portNumber.setText("");
return;
}
ChatServer.port = savePort;
dispose();
} catch (NumberFormatException e) {
PortConf.DLGINFO.setText("错误的端口号,端口号请填写整数!");
PortConf.portNumber.setText("");
return;
}
}
});
//关闭对话框时的操作
this.addWindowListener(new WindowAdapter() {
public void windowClosing(WindowEvent e) {DLGINFO.setText("默认端口号为:8888");
}
});
//取消按钮的事件处理
cancel.addActionListener(new ActionListener() {
public void actionPerformed(ActionEvent e) {DLGINFO.setText("默认端口号为:8888");
dispose();
}
});
}
}
```

5. Node.java

```java
import java.net.*;
import java.io.*;
/**
*用户链表的结点类
*/
public class Node {
String username = null;
Socket socket = null;
ObjectOutputStream output = null;
ObjectInputStream input = null;
```

```java
Node next = null;
}
```

6. userLinklist.java

```java
/**
 * 用户链表
 */
public class UserLinkList {
    Node root;
    Node pointer;
    int count;
    /**
     * 构造用户链表
     */
    public UserLinkList() {
        root = new Node();
        root.next = null;
        pointer = null;
        count = 0;
    }
    /**
     * 添加用户
     */
    public void addUser(Node n) {
        pointer = root;
        while (pointer.next != null) {
            pointer = pointer.next;
        }
        pointer.next = n;
        n.next = null;
        count++;
    }
    /**
     * 删除用户
     */
    public void delUser(Node n) {
        pointer = root;
        while (pointer.next != null) {
            if (pointer.next == n) {
                pointer.next = n.next;
                count--;
                break;
            }
            pointer = pointer.next;
        }
```

```java
}
/**
 * 返回用户数
 */
public int getCount() {
    return count;
}
/**
 * 根据用户名查找用户
 */
public Node findUser(String username) {
    if (count==0)
        return null;
    pointer=root;
    while (pointer.next!=null) {
        pointer=pointer.next;
        if (pointer.username.equalsIgnoreCase(username)) {
            return pointer;
        }
    }
    return null;
}
/**
 * 根据索引查找用户
 */
public Node findUser(int index) {
    if (count==0) {
        return null;
    }
    if (index < 0) {
        return null;
    }
    pointer=root;
    int i=0;
    while (i < index+1) {
        if (pointer.next!=null) {
            pointer=pointer.next;
        } else {
            return null;
        }
        i++;
    }
    return pointer;
}
```

}

7. help.java

```java
import java.awt.*;
import javax.swing.border.*;
import java.net.*;
import javax.swing.*;
import java.awt.event.*;
/**
 *生成设置对话框的类
 */
public class Help extends JDialog {
JPanel titlePanel = new JPanel();
JPanel contentPanel = new JPanel();
JPanel closePanel = new JPanel();
JButton close = new JButton();
JLabel title = new JLabel("聊天室服务端帮助");
JTextArea help = new JTextArea();
Color bg = new Color(255, 255, 255);
public Help(JFrame frame) {
super(frame, true);
try {
jbInit();
} catch (Exception e) {
e.printStackTrace();
}
//设置运行位置,使对话框居中
Dimension screenSize = Toolkit.getDefaultToolkit().getScreenSize();
this.setLocation((int)(screenSize.width - 400) / 2,
(int)(screenSize.height - 320) / 2);
this.setResizable(false);
}
private void jbInit() throws Exception {
this.setSize(new Dimension(400, 200));
this.setTitle("帮助");
titlePanel.setBackground(bg);
;
contentPanel.setBackground(bg);
closePanel.setBackground(bg);
help.setText("1、设置服务端的侦听端口(默认端口为8888)。\n"
+ "2、点击 启动服务 按钮便可在指定的端口启动服务。\n"
+ "3、选择需要接受消息的用户,在消息栏中写入消息,之后便可发送消息。\n"
+ "4、信息状态栏中显示服务器当前的启动与停止状态、"+"用户发送的消息和服务器端发送的系统消息。");
help.setEditable(false);
```

```java
titlePanel.add(new Label(""));
titlePanel.add(title);
titlePanel.add(new Label(""));
contentPanel.add(help);
closePanel.add(new Label(""));
closePanel.add(close);
closePanel.add(new Label(""));
Container contentPane=getContentPane();
contentPane.setLayout(new BorderLayout());
contentPane.add(titlePanel,BorderLayout.NORTH);
contentPane.add(contentPanel,BorderLayout.CENTER);
contentPane.add(closePanel,BorderLayout.SOUTH);
close.setText("关闭");
//事件处理
close.addActionListener(new ActionListener(){
    public void actionPerformed(ActionEvent e){
        dispose();
    }
});
}
}
```

客户端代码:
1. chantclent.java

```java
import java.awt.*;
import java.awt.event.*;
import javax.swing.*;
import javax.swing.event.*;
import java.io.*;
import java.net.*;

/*
 *聊天客户端的主框架类
 */
public class ChatClient extends JFrame implements ActionListener{
    String ip="127.0.0.1";//连接到服务端的ip地址
    int port=8888;// 连接到服务端的端口号
    String userName="匆匆过客";// 用户名
    int type=0;// 0 表示未连接,1 表示已连接
    Image icon;//程序图标
    JComboBox combobox;//选择发送消息的接受者
    JTextArea messageShow;//客户端的信息显示
    JScrollPane messageScrollPane;//信息显示的滚动条
    JLabel express, sendToLabel, messageLabel;
    JTextField clientMessage;//客户端消息的发送
```

```java
JCheckBox checkbox;//悄悄话
JComboBox actionlist;//表情选择
JButton clientMessageButton;//发送消息
JTextField showStatus;//显示用户连接状态
Socket socket;
ObjectOutputStream output;//网络套接字输出流
ObjectInputStream input;//网络套接字输入流
ClientReceive recvThread;
//建立菜单栏
JMenuBar jMenuBar = new JMenuBar( );
//建立菜单组
JMenu operateMenu = new JMenu("操作(O)");
//建立菜单项
JMenuItem loginItem = new JMenuItem("用户登录(I)");
JMenuItem logoffItem = new JMenuItem("用户注销(L)");
JMenuItem exitItem = new JMenuItem("退出(X)");
JMenu conMenu = new JMenu("设置(C)");
JMenuItem userItem = new JMenuItem("用户设置(U)");
JMenuItem connectItem = new JMenuItem("连接设置(C)");
JMenu helpMenu = new JMenu("帮助(H)");
JMenuItem helpItem = new JMenuItem("帮助(H)");
//建立工具栏
JToolBar toolBar = new JToolBar( );
//建立工具栏中的按钮组件
JButton loginButton;//用户登录
JButton logoffButton;//用户注销
JButton userButton;//用户信息的设置
JButton connectButton;//连接设置
JButton exitButton;//退出按钮
//框架的大小
Dimension faceSize = new Dimension(400, 600);
JPanel downPanel;
GridBagLayout girdBag;
GridBagConstraints girdBagCon;
public ChatClient( ) {
init( );// 初始化程序
// 添加框架的关闭事件处理
this.setDefaultCloseOperation(JFrame.EXIT_ON_CLOSE);
this.pack( );
// 设置框架的大小
this.setSize(faceSize);
// 设置运行时窗口的位置
Dimension screenSize = Toolkit.getDefaultToolkit( ).getScreenSize( );
this.setLocation((int) (screenSize.width - faceSize.getWidth( )) / 2,
```

```
(int)(screenSize.height – faceSize.getHeight())/2);
this.setResizable(false);
this.setTitle("聊天室客户端");// 设置标题
// 程序图标
icon = getImage("icon.gif");
this.setIconImage(icon);// 设置程序图标
show();
// 为操作菜单栏设置热键'V'
operateMenu.setMnemonic('O');
// 为用户登录设置快捷键为 ctrl+i
loginItem.setMnemonic('I');
loginItem.setAccelerator(KeyStroke.getKeyStroke(KeyEvent.VK_I,
InputEvent.CTRL_MASK));
// 为用户注销快捷键为 ctrl+l
logoffItem.setMnemonic('L');
logoffItem.setAccelerator(KeyStroke.getKeyStroke(KeyEvent.VK_L,
InputEvent.CTRL_MASK));
// 为退出快捷键为 ctrl+x
exitItem.setMnemonic('X');
exitItem.setAccelerator(KeyStroke.getKeyStroke(KeyEvent.VK_X,
InputEvent.CTRL_MASK));
// 为设置菜单栏设置热键'C'
conMenu.setMnemonic('C');
// 为用户设置设置快捷键为 ctrl+u
userItem.setMnemonic('U');
userItem.setAccelerator(KeyStroke.getKeyStroke(KeyEvent.VK_U,
InputEvent.CTRL_MASK));
// 为连接设置设置快捷键为 ctrl+c
connectItem.setMnemonic('C');
connectItem.setAccelerator(KeyStroke.getKeyStroke(KeyEvent.VK_C,
InputEvent.CTRL_MASK));
// 为帮助菜单栏设置热键'H'
helpMenu.setMnemonic('H');
// 为帮助设置快捷键为 ctrl+p
helpItem.setMnemonic('H');
helpItem.setAccelerator(KeyStroke.getKeyStroke(KeyEvent.VK_H,
InputEvent.CTRL_MASK));
}
/**
 * 程序初始化函数
 */
public void init() {
Container contentPane = getContentPane();
contentPane.setLayout(new BorderLayout());
```

```
// 添加菜单栏
operateMenu.add(loginItem);
operateMenu.add(logoffItem);
operateMenu.add(exitItem);
jMenuBar.add(operateMenu);
conMenu.add(userItem);
conMenu.add(connectItem);
jMenuBar.add(conMenu);
helpMenu.add(helpItem);
jMenuBar.add(helpMenu);
setJMenuBar(jMenuBar);
// 初始化按钮
loginButton = new JButton("登录");
logoffButton = new JButton("注销");
userButton = new JButton("用户设置");
connectButton = new JButton("连接设置");
exitButton = new JButton("退出");
// 当鼠标放上显示信息
loginButton.setToolTipText("连接到指定的服务器");
logoffButton.setToolTipText("与服务器断开连接");
userButton.setToolTipText("设置用户信息");
connectButton.setToolTipText("设置所要连接到的服务器信息");
// 将按钮添加到工具栏
toolBar.add(userButton);
toolBar.add(connectButton);
toolBar.addSeparator();// 添加分隔栏
toolBar.add(loginButton);
toolBar.add(logoffButton);
toolBar.addSeparator();// 添加分隔栏
toolBar.add(exitButton);
contentPane.add(toolBar, BorderLayout.NORTH);
checkbox = new JCheckBox("悄悄话");
checkbox.setSelected(false);
actionlist = new JComboBox();
actionlist.addItem("微笑地");
actionlist.addItem("高兴地");
actionlist.addItem("轻轻地");
actionlist.addItem("生气地");
actionlist.addItem("小心地");
actionlist.addItem("静静地");
actionlist.setSelectedIndex(0);
// 初始时
loginButton.setEnabled(true);
logoffButton.setEnabled(false);
```

```java
// 为菜单栏添加事件监听
loginItem.addActionListener(this);
logoffItem.addActionListener(this);
exitItem.addActionListener(this);
userItem.addActionListener(this);
connectItem.addActionListener(this);
helpItem.addActionListener(this);
// 添加按钮的事件侦听
loginButton.addActionListener(this);
logoffButton.addActionListener(this);
userButton.addActionListener(this);
connectButton.addActionListener(this);
exitButton.addActionListener(this);
combobox = new JComboBox();
combobox.insertItemAt("所有人", 0);
combobox.setSelectedIndex(0);
messageShow = new JTextArea();
messageShow.setEditable(false);
// 添加滚动条
messageScrollPane = new JScrollPane(messageShow,
    JScrollPane.VERTICAL_SCROLLBAR_AS_NEEDED,
    JScrollPane.HORIZONTAL_SCROLLBAR_AS_NEEDED);
messageScrollPane.setPreferredSize(new Dimension(400, 400));
messageScrollPane.revalidate();
clientMessage = new JTextField(23);
clientMessage.setEnabled(false);
clientMessageButton = new JButton();
clientMessageButton.setText("发送");
// 添加系统消息的事件侦听
clientMessage.addActionListener(this);
clientMessageButton.addActionListener(this);
sendToLabel = new JLabel("发送至:");
express = new JLabel("表情:");
messageLabel = new JLabel("发送消息:");
downPanel = new JPanel();
girdBag = new GridBagLayout();
downPanel.setLayout(girdBag);
girdBagCon = new GridBagConstraints();
girdBagCon.gridx = 0;
girdBagCon.gridy = 0;
girdBagCon.gridwidth = 5;
girdBagCon.gridheight = 2;
girdBagCon.ipadx = 5;
girdBagCon.ipady = 5;
```

```java
JLabel none = new JLabel("    ");
girdBag.setConstraints(none, girdBagCon);
downPanel.add(none);
girdBagCon = new GridBagConstraints();
girdBagCon.gridx = 0;
girdBagCon.gridy = 2;
girdBagCon.insets = new Insets(1, 0, 0, 0);
// girdBagCon.ipadx = 5;
// girdBagCon.ipady = 5;
girdBag.setConstraints(sendToLabel, girdBagCon);
downPanel.add(sendToLabel);
girdBagCon = new GridBagConstraints();
girdBagCon.gridx = 1;
girdBagCon.gridy = 2;
girdBagCon.anchor = GridBagConstraints.LINE_START;
girdBag.setConstraints(combobox, girdBagCon);
downPanel.add(combobox);
girdBagCon = new GridBagConstraints();
girdBagCon.gridx = 2;
girdBagCon.gridy = 2;
girdBagCon.anchor = GridBagConstraints.LINE_END;
girdBag.setConstraints(express, girdBagCon);
downPanel.add(express);
girdBagCon = new GridBagConstraints();
girdBagCon.gridx = 3;
girdBagCon.gridy = 2;
girdBagCon.anchor = GridBagConstraints.LINE_START;
// girdBagCon.insets = new Insets(1,0,0,0);
// girdBagCon.ipadx = 5;
// girdBagCon.ipady = 5;
girdBag.setConstraints(actionlist, girdBagCon);
downPanel.add(actionlist);
girdBagCon = new GridBagConstraints();
girdBagCon.gridx = 4;
girdBagCon.gridy = 2;
girdBagCon.insets = new Insets(1, 0, 0, 0);
// girdBagCon.ipadx = 5;
// girdBagCon.ipady = 5;
girdBag.setConstraints(checkbox, girdBagCon);
downPanel.add(checkbox);
girdBagCon = new GridBagConstraints();
girdBagCon.gridx = 0;
girdBagCon.gridy = 3;
girdBag.setConstraints(messageLabel, girdBagCon);
```

```java
downPanel.add(messageLabel);
girdBagCon = new GridBagConstraints();
girdBagCon.gridx = 1;
girdBagCon.gridy = 3;
girdBagCon.gridwidth = 3;
girdBagCon.gridheight = 1;
girdBag.setConstraints(clientMessage, girdBagCon);
downPanel.add(clientMessage);
girdBagCon = new GridBagConstraints();
girdBagCon.gridx = 4;
girdBagCon.gridy = 3;
girdBag.setConstraints(clientMessageButton, girdBagCon);
downPanel.add(clientMessageButton);
showStatus = new JTextField(35);
showStatus.setEditable(false);
girdBagCon = new GridBagConstraints();
girdBagCon.gridx = 0;
girdBagCon.gridy = 5;
girdBagCon.gridwidth = 5;
girdBag.setConstraints(showStatus, girdBagCon);
downPanel.add(showStatus);
contentPane.add(messageScrollPane, BorderLayout.CENTER);
contentPane.add(downPanel, BorderLayout.SOUTH);
// 关闭程序时的操作
this.addWindowListener(new WindowAdapter() {
    public void windowClosing(WindowEvent e) {
        if (type == 1) {
            DisConnect();
        }
        System.exit(0);
    }
});
}
/**
 * 事件处理
 */
public void actionPerformed(ActionEvent e) {
    Object obj = e.getSource();
    if (obj == userItem || obj == userButton) { // 用户信息设置
        // 调出用户信息设置对话框
        UserConf userConf = new UserConf(this, userName);
        userConf.show();
        userName = userConf.userInputName;
    } else if (obj == connectItem || obj == connectButton) { // 连接服务端设置
```

```java
// 调出连接设置对话框
ConnectConf conConf=new ConnectConf(this, ip, port);
conConf.show();
ip=conConf.userInputIp;
port=conConf.userInputPort;
} else if (obj==loginItem || obj==loginButton) { // 登录
Connect();
} else if (obj==logoffItem || obj==logoffButton) { // 注销
DisConnect();
showStatus.setText("");
} else if (obj==clientMessage || obj==clientMessageButton) { // 发送消息
SendMessage();
clientMessage.setText("");
} else if (obj==exitButton || obj==exitItem) { // 退出
int j=JOptionPane.showConfirmDialog(this,"真的要退出吗?","退出",
JOptionPane.YES_OPTION, JOptionPane.QUESTION_MESSAGE);
if (j==JOptionPane.YES_OPTION) {
if (type==1) {
DisConnect();
}
System.exit(0);
}
} else if (obj==helpItem) { // 菜单栏中的帮助
// 调出帮助对话框
Help helpDialog=new Help(this);
helpDialog.show();
}
}
public void Connect() {
try {
socket=new Socket(ip, port);
} catch (Exception e) {
JOptionPane.showConfirmDialog(this,"不能连接到指定的服务器。\n请确认连接设置是否正确。",
"提示", JOptionPane.DEFAULT_OPTION,JOptionPane.WARNING_MESSAGE);
return;
}
try {
output=new ObjectOutputStream(socket.getOutputStream());
output.flush();
input=new ObjectInputStream(socket.getInputStream());
output.writeObject(userName);
output.flush();
recvThread=new ClientReceive(socket, output, input, combobox,
messageShow, showStatus);
```

```java
recvThread.start();
loginButton.setEnabled(false);
loginItem.setEnabled(false);
userButton.setEnabled(false);
userItem.setEnabled(false);
connectButton.setEnabled(false);
connectItem.setEnabled(false);
logoffButton.setEnabled(true);
logoffItem.setEnabled(true);
clientMessage.setEnabled(true);
messageShow.append("连接服务器 "+ip+":"+port+" 成功...\n");
type=1;// 标志位设为已连接
} catch (Exception e) {
System.out.println(e);
return;
}
}
public void DisConnect() {
loginButton.setEnabled(true);
loginItem.setEnabled(true);
userButton.setEnabled(true);
userItem.setEnabled(true);
connectButton.setEnabled(true);
connectItem.setEnabled(true);
logoffButton.setEnabled(false);
logoffItem.setEnabled(false);
clientMessage.setEnabled(false);
if (socket.isClosed()) {
return;
}
try {
output.writeObject("用户下线");
output.flush();
input.close();
output.close();
socket.close();
messageShow.append("已经与服务器断开连接...\n");
type=0;// 标志位设为未连接
} catch (Exception e) {
}
}
public void SendMessage() {
String toSomebody=combobox.getSelectedItem().toString();
String status="";
```

```java
if (checkbox.isSelected()) {
status = "悄悄话";
}
String action = actionlist.getSelectedItem().toString();
String message = clientMessage.getText();
if (socket.isClosed()) {
return;
}
try {
output.writeObject("聊天信息");
output.flush();
output.writeObject(toSomebody);
output.flush();
output.writeObject(status);
output.flush();
output.writeObject(action);
output.flush();
output.writeObject(message);
output.flush();
} catch (Exception e) {
}
}
/**
 * 通过给定的文件名获得图像
 */
Image getImage(String filename) {
URLClassLoader urlLoader = (URLClassLoader) this.getClass()
    .getClassLoader();
URL url = null;
Image image = null;
url = urlLoader.findResource(filename);
image = Toolkit.getDefaultToolkit().getImage(url);
MediaTracker mediatracker = new MediaTracker(this);
try {
mediatracker.addImage(image, 0);
mediatracker.waitForID(0);
} catch (InterruptedException ex) {
image = null;
}
if (mediatracker.isErrorID(0)) {
image = null;
}
return image;
}
```

```java
public static void main(String[] args) {
ChatClient app = new ChatClient();
}
}
```

8. clentreceive.java

```java
import javax.swing.*;
import java.io.*;
import java.net.*;
/*
*聊天客户端消息收发类
*/
public class ClientReceive extends Thread {
private JComboBox combobox;
private JTextArea textarea;
Socket socket;
ObjectOutputStream output;
ObjectInputStream input;
JTextField showStatus;
public ClientReceive(Socket socket, ObjectOutputStream output,
ObjectInputStream input, JComboBox combobox, JTextArea textarea,
JTextField showStatus) {
this.socket = socket;
this.output = output;
this.input = input;
this.combobox = combobox;
this.textarea = textarea;
this.showStatus = showStatus;
}
public void run() {
while (!socket.isClosed()) {
try {
String type = (String) input.readObject();
if (type.equalsIgnoreCase("系统信息")) {
String sysmsg = (String) input.readObject();
textarea.append("系统信息："+sysmsg);
} else if (type.equalsIgnoreCase("服务关闭")) {
output.close();
input.close();
socket.close();
textarea.append("服务器已关闭！\n");
break;
} else if (type.equalsIgnoreCase("聊天信息")) {
String message = (String) input.readObject();
textarea.append(message);
```

```java
} else if (type.equalsIgnoreCase("用户列表")) {
String userlist = (String) input.readObject();
String usernames[] = userlist.split("\n");
combobox.removeAllItems();
int i = 0;
combobox.addItem("所有人");
while (i < usernames.length) {
combobox.addItem(usernames[i]);
i++;
}
combobox.setSelectedIndex(0);
showStatus.setText("在线用户 "+usernames.length+" 人");
}
} catch (Exception e) {
System.out.println(e);
}
}
}
}
```

9. connectconf.java

```java
import java.awt.*;
import javax.swing.border.*;
import java.net.*;
import javax.swing.*;
import java.awt.event.*;
/**
 *生成连接信息输入的对话框
 *让用户输入连接服务器的 IP 和端口
 */
public class ConnectConf extends JDialog {
JPanel panelUserConf = new JPanel();
JButton save = new JButton();
JButton cancel = new JButton();
JLabel DLGINFO = new JLabel("默认连接设置为 127.0.0.1:8888");
JPanel panelSave = new JPanel();
JLabel message = new JLabel();
String userInputIp;
int userInputPort;
JTextField inputIp;
JTextField inputPort;
public ConnectConf(JFrame frame, String ip, int port) {
super(frame, true);
this.userInputIp = ip;
this.userInputPort = port;
```

```
try {
    jbInit();
}
catch (Exception e) {
    e.printStackTrace();
}
//设置运行位置,使对话框居中
Dimension screenSize = Toolkit.getDefaultToolkit().getScreenSize();
this.setLocation((int)(screenSize.width - 400)/2+50,
    (int)(screenSize.height - 600)/2+150);
this.setResizable(false);
}
private void jbInit() throws Exception {
    this.setSize(new Dimension(300,130));
    this.setTitle("连接设置");
    message.setText("请输入服务器的IP地址:");
    inputIp = new JTextField(10);
    inputIp.setText(userInputIp);
    inputPort = new JTextField(4);
    inputPort.setText(""+userInputPort);
    save.setText("保存");
    cancel.setText("取消");
    panelUserConf.setLayout(new GridLayout(2,2,1,1));
    panelUserConf.add(message);
    panelUserConf.add(inputIp);
    panelUserConf.add(new JLabel("请输入服务器的端口号:"));
    panelUserConf.add(inputPort);
    panelSave.add(new Label(""));
    panelSave.add(save);
    panelSave.add(cancel);
    panelSave.add(new Label(""));
    Container contentPane = getContentPane();
    contentPane.setLayout(new BorderLayout());
    contentPane.add(panelUserConf, BorderLayout.NORTH);
    contentPane.add(DLGINFO, BorderLayout.CENTER);
    contentPane.add(panelSave, BorderLayout.SOUTH);
    //保存按钮的事件处理
    save.addActionListener(new ActionListener() {
        public void actionPerformed(ActionEvent a) {
            int savePort;
            String inputIP;
            //判断端口号是否合法
            try {
                userInputIp = ""+InetAddress.getByName(inputIp.getText());
                userInputIp = userInputIp.substring(1);
```

```java
} catch (UnknownHostException e) {
DLGINFO
.setText("错误的 IP 地址!");
return;
}
//userInputIp = inputIP;
//判断端口号是否合法
try {
savePort = Integer.parseInt(inputPort.getText());
if (savePort < 1 || savePort > 65535) {
DLGINFO.setText("侦听端口必须是 0-65535 之间的整数!");
inputPort.setText("");
return;
}
userInputPort = savePort;
dispose();
} catch (NumberFormatException e) {
DLGINFO.setText("错误的端口号,端口号请填写整数!");
inputPort.setText("");
return;
}
}
});
//关闭对话框时的操作
this.addWindowListener(new WindowAdapter() {
public void windowClosing(WindowEvent e) {
DLGINFO.setText("默认连接设置为 127.0.0.1:8888");
}
});
//取消按钮的事件处理
cancel.addActionListener(new ActionListener() {
public void actionPerformed(ActionEvent e) {
DLGINFO.setText("默认连接设置为 127.0.0.1:8888");
dispose();
}
});
}
}
```

10. userconf.java

```java
import java.awt.*;
import javax.swing.border.*;
import java.net.*;
import javax.swing.*;
import java.awt.event.*;
```

```
/**
*生成用户信息输入对话框的类
*让用户输入自己的用户名
*/
public class UserConf extends JDialog {
JPanel panelUserConf=new JPanel();
JButton save=new JButton();
JButton cancel=new JButton();
JLabel DLGINFO=new JLabel("默认用户名为:匆匆过客");
JPanel panelSave=new JPanel();
JLabel message=new JLabel();
String userInputName;
JTextField userName;
public UserConf(JFrame frame, String str) {
super(frame, true);
this.userInputName=str;
try {
jbInit();
} catch (Exception e) {
e.printStackTrace();
}
//设置运行位置,使对话框居中
Dimension screenSize=Toolkit.getDefaultToolkit().getScreenSize();
this.setLocation((int)(screenSize.width - 400)/2+50,
(int)(screenSize.height - 600)/2+150);
this.setResizable(false);
}
private void jbInit() throws Exception {
this.setSize(new Dimension(300, 120));
this.setTitle("用户设置");
message.setText("请输入用户名:");
userName=new JTextField(10);
userName.setText(userInputName);
save.setText("保存");
cancel.setText("取消");
panelUserConf.setLayout(new FlowLayout());
panelUserConf.add(message);
panelUserConf.add(userName);
panelSave.add(new Label(""));
panelSave.add(save);
panelSave.add(cancel);
panelSave.add(new Label(""));
Container contentPane=getContentPane();
contentPane.setLayout(new BorderLayout());
```

```java
contentPane.add(panelUserConf, BorderLayout.NORTH);
contentPane.add(DLGINFO, BorderLayout.CENTER);
contentPane.add(panelSave, BorderLayout.SOUTH);
//保存按钮的事件处理
save.addActionListener(new ActionListener() {
    public void actionPerformed(ActionEvent a) {
        if (userName.getText().equals("")) {
            DLGINFO
            .setText("用户名不能为空!");
            userName.setText(userInputName);
            return;
        } else if (userName.getText().length() > 15) {
            DLGINFO.setText("用户名长度不能大于15个字符!");
            userName.setText(userInputName);
            return;
        }
        userInputName = userName.getText();
        dispose();
    }
});
//关闭对话框时的操作
this.addWindowListener(new WindowAdapter() {
    public void windowClosing(WindowEvent e) {
        DLGINFO.setText("默认用户名为:匆匆过客");
    }
});
//取消按钮的事件处理
cancel.addActionListener(new ActionListener() {
    public void actionPerformed(ActionEvent e) {
        DLGINFO.setText("默认用户名为:匆匆过客");
        dispose();
    }
});
    }
}
```

11. help.java

```java
import java.awt.*;
import javax.swing.border.*;
import java.net.*;
import javax.swing.*;
import java.awt.event.*;
/**
 *生成设置对话框的类
 */
```

```java
public class Help extends JDialog {
    JPanel titlePanel = new JPanel();
    JPanel contentPanel = new JPanel();
    JPanel closePanel = new JPanel();
    JButton close = new JButton();
    JLabel title = new JLabel("聊天室客户端帮助");
    JTextArea help = new JTextArea();
    Color bg = new Color(255, 255, 255);
    public Help(JFrame frame) {
        super(frame, true);
        try {
            jbInit();
        } catch (Exception e) {
            e.printStackTrace();
        }
        //设置运行位置,使对话框居中
        Dimension screenSize = Toolkit.getDefaultToolkit().getScreenSize();
        this.setLocation((int)(screenSize.width - 400) / 2 + 25,
            (int)(screenSize.height - 320) / 2);
        this.setResizable(false);
    }
    private void jbInit() throws Exception {
        this.setSize(new Dimension(350, 270));
        this.setTitle("帮助");
        titlePanel.setBackground(bg);
        contentPanel.setBackground(bg);
        closePanel.setBackground(bg);
        help.setText("1、设置所要连接服务端的 IP 地址和端口"+"(默认设置为\n        127.0.0.1:8888)。\n"
            + "2、输入你的用户名(默认设置为:匆匆过客)。\n"+"3、点击"登录"便可以连接到指定的服务器;\n"
            + "        点击"注销"可以和服务器端开连接。\n"+"4、选择需要接受消息的用户,在消息栏中写入消息,\n"
            + "        同时选择表情,之后便可发送消息。\n");
        help.setEditable(false);
        titlePanel.add(new Label(""));
        titlePanel.add(title);
        titlePanel.add(new Label(""));
        contentPanel.add(help);
        closePanel.add(new Label(""));
        closePanel.add(close);
        closePanel.add(new Label(""));
        Container contentPane = getContentPane();
        contentPane.setLayout(new BorderLayout());
        contentPane.add(titlePanel, BorderLayout.NORTH);
        contentPane.add(contentPanel, BorderLayout.CENTER);
```

```
contentPane.add(closePanel, BorderLayout.SOUTH);
close.setText("关闭");
//事件处理
close.addActionListener(new ActionListener() {
    public void actionPerformed(ActionEvent e) {
        dispose();
    }
});
}
}
```

四、程序的运行与发布

1. 聊天室程序运行

将 ChatServer.java、ServerListen.java、ServerReceive.java、PortConf.java、Node.java、UserLinkList.java 和 Help.java 这 7 个文件保存到一个文件夹中。利用 javac 命令对文件进行编辑，使用的命令如下：

javac ChatServer.java

之后，利用 java 命令执行程序，使用的命令如下：

javaChatServer

2. 聊天室服务器端程序发布

要发布此应用程序，需要将应用程序打包。使用 jar.exe 可以把应用程序中涉及文件和图片压缩成一个 jar 文件，这样便可以发布程序。

首先编写一个清单文件，名为 MANIFEST.MF，其代码如下：

Manifest-Version: 1.0

Created-By: 1.5.0_02(Sun microsystem Inc.)

Main-class: ChatServer

将此清单文件保存到 C:\Javawork\ch08 文件夹中。

然后，使用如下命令生成 jar 文件：

jar cfmChatServer.jar MANIFEST.MF *.class

其中参数 c 表示要生成一个新的 jar 文件；f 表示要生成的 jar 文件的名字；m 表示清单文件的名字。

如果机器安装这 WinRAR 解压软件，并将.jar 文件与解压缩软件做了关联，那么 ChatServer.jar 文件的类型是 WinRAR，使得 Java 程序无法运行。因此，我们在发布软件时，还应该再写一个有如下内容的 bat 文件（ChatServer.bat）：

javaw - jar ChatServer.jar

然后可以通过双击 Spider.bat 来运行程序。

3. 聊天室客户端程序运行

将 ChatClient.java、ClientReceive.java、ConnectConf.java、UserConf.java 和 Help.java 这 5 个文件保存到一个文件夹中。利用 javac 命令对文件进行编辑，使用的命令如下：

javac ChatClient.java

之后，利用 java 命令执行程序，使用的命令如下：

javaChatClient

4. 聊天室客户端程序发布

要发布此应用程序，需要将应用程序打包。使用 jar.exe，可以把应用程序中涉及文件和图片压缩成一个 jar 文件，这样便可以发布程序。

首先编写一个清单文件，名为 MANIFEST.MF，其代码如下：

Manifest-Version：1.0

Created-By：1.6.0(Sun microsystem Inc.)

Main-class：ChatClient

将此清单文件保存到 C:\Javawork\ch08 文件夹中。

然后，使用如下命令生成 jar 文件：

jar cfm ChatServer.jar MANIFEST.MF *.class

其中参数 c 表示要生成一个新的 jar 文件；f 表示要生成的 jar 文件的名字；m 表示清单文件的名字。

如果机器安装这 WinRAR 解压软件，并将.jar 文件与解压缩软件做了关联，那么 ChatClient.jar 文件的类型是 WinRAR，使得 Java 程序无法运行。因此，我们在发布软件时，还应该再写一个有如下内容的 bat 文件(ChatClient.bat)：

javaw - jar ChatClient.jar

然后可以通过双击 ChatClient.bat 来运行程序。

参考文献

[1] 朱庆生. Java 程序设计之实验及课程设计教程[M]. 北京:清华大学出版社,2011.
[2] 李兆锋,庞永庆. Java 程序设计与项目实践[M]. 北京:电子工业出版社,2011.
[3] 张跃平,耿祥义. Java 程序设计精编教程实验指导与习题解答[M]. 北京:清华大学出版社,2012.
[4] 王占中. Java 程序开发基础教程与实验指导[M]. 北京:清华大学出版社,2012.
[5] 邹林达,陈国君. Java 程序设计基础实验指导[M]. 3 版. 北京:清华大学出版社,2011.
[6] 施珺,纪光辉. JAVA 语言实验与课程设计指导[M]. 南京:南京大学出版社,2010.
[7] 丁振凡. Java 语言程序设计实验指导与习题解答[M]. 北京:清华大学出版社,2010.
[8] 陈铁,姚晓昆. 新编 Java 程序设计实验指导[M]. 北京:清华大学出版社,2010.
[9] 赵欢. Java 程序设计实用教程实验指导、实训与习题解析[M]. 北京:水利水电出版社,2009.
[10] 范玫,马俊. Java 语言面向对象程序设计实验指导与习题问答[M]. 北京:机械工业出版社,2009.
[11] 飞思科技产品研发中心. 精通 Jbuilder9[M]. 北京:电子工业出版社,2009.
[12] 袁海燕,王文涛. JAVA 实用程序设计 100 例[M]. 北京:人民邮电出版社,2009.
[13] 赵文靖. Java 程序设计基础与上机指导[M]. 北京:清华大学出版社,2006.
[14] 张屹,蔡木生. Java 核心编程技术实验指导教程[M]. 大连:大连理工大学出版社,2010.
[15] 雍俊海. Java 程序设计习题集[M]. 北京:清华大学出版社,2006.
[16] 朱福喜. Java 语言习题与解析[M]. 北京:清华大学出版社,2006.
[17] 吴其庆. Java 程序设计实例教程[M]. 北京:冶金工业出版社,2006.
[18] 孙卫琴,李洪成. Tomcat 与 Java Web 开发技术详解[M]. 北京:电子工业出版社,2003.
[19] 孙一林,彭波. Java 数据库编程实例[M]. 北京:清华大学出版社,2002.
[20] 赛奎春. JAVA 工程应用与项目实践[M]. 北京:机械工业出版社,2002.
[21] 郎波. Java 语言程序设计[M]. 2 版. 北京:清华大学出版社,2010.
[22] 唐振明. Java 程序设计[M]. 北京:电子工业出版社,2011.
[23] 张晓龙. Java 程序设计与开发[M]. 北京:电子工业出版社,2010.
[24] 常建功. 零基础学 Java[M]. 3 版. 北京:电子工业出版社,2012.
[25] ROGERS CADENHEAD. Java 入门经典[M]. 6 版. 梅兴文,郝记生,译. 北京:人民邮电出版社,2012.
[26] BRUCEECKEL. Java 编程思想[M]. 北京:机械工业出版社,2003.
[27] FLANAGAN. Java 技术手册[M]. 北京:中国电力出版社,2002.
[28] Brown,等. JAVA 编程指南[M]. 2 版. 北京:电子工业出版社,2003.
[29] 李尊朝,苏军. Java 语言程序设计例题解析与实验指导[M]. 2 版. 北京:中国铁道出版社,2008.
[30] 刘宝林. Java 程序设计与案例习题解答与实验指导[M]. 北京:高等教育出版社,2005.
[31] 余青松,江红. Java 程序设计实验指导与习题测试[M]. 北京:清华大学出版社,2012.